SpringerBriefs in Earth System Sciences

SpringerBriefs South America and the Southern Hemisphere

Series Editors

Gerrit Lohmann
Lawrence A. Mysak
Justus Notholt
Jorge Rabassa
Vikram Unnithan

For further volumes:
http://www.springer.com/series/10032

Federico L. Agnolín · Fernando E. Novas

Avian Ancestors

A Review of the Phylogenetic Relationships
of the Theropods Unenlagiidae,
Microraptoria, *Anchiornis*
and Scansoriopterygidae

 Springer

Federico L. Agnolín
"Félix de Azara", Departamento de
 Ciencias Naturales
Fundación de Historia Natural,
 CEBBAD, Universidad Maimónides
Buenos Aires
Argentina

Fernando E. Novas
CONICET, Museo Argentino de Ciencias Naturales
 "Bernardino Rivadavia"
Buenos Aires
Argentina

ISSN 2191-589X   ISSN 2191-5903   (electronic)
ISBN 978-94-007-5636-6 ISBN 978-94-007-5637-3   (eBook)
DOI 10.1007/978-94-007-5637-3
Springer Dordrecht Heidelberg New York London

Library of Congress Control Number: 2012953463

Printed on acid-free paper

Springer is part of Springer Science+Business Media (www.springer.com)

Preface

Although consensus exists among authors that birds evolved from coelurosaurian theropods, paleontologists still debate the identification of the group of coelurosaurians that most closely approaches the common ancestor of birds. The past 20 years witnessed the discovery of a wide array of avian-like theropods that has considerably amplified the anatomical disparity among deinonychosaurians, some of which resemble *Archaeopteryx* more than *Deinonychus*. Among these newly discovered theropods that show remarkable bird-like characteristics are the four-winged theropods *Microraptor* and *Anchiornis*, and the unenlagiids *Unenlagia, Buitreraptor,* and *Rahonavis*. *Xiaotingia, Anchiornis*, and *Archaeopteryx* are regarded as more nearly related to birds, rather than to Dromaeosauridae or Troodontidae. Moreover, a bizarre group of minute-sized coelurosaurs, the Scansoriopterygidae, also exhibits some avian similarities that lead some authors to interpret them as more closely related to birds than other dinosaurs. With the aim to explore the phylogenetic relationships of these coelurosaurians and birds, we merged recently published integrative databases, resulting in significant changes in the topological distribution of taxa within Paraves. We present evidence that Dromaeosauridae, Microraptoria, Unenlagiidae, and *Anchiornis + Xiaotingia* form successive sister taxa of Aves, and that the Scansoriopterygidae are basal coelurosaurians not closely related to birds. The implications in the evolutionary sequence of anatomical characters leading to birds, including the origin of flight, are also considered in light of this new phylogenetic hypothesis.

Keywords Microraptoria • Unenlagiidae • *Anchiornis* • Scansoriopterygidae • Origin of birds and flight

Acknowledgments

We thank G. Mayr for the information about the 10th *Archaeopteryx* skeleton. We acknowledge F. Gianechini, R. Lucero, and M. Ezcurra for fruitful discussions about theropod phylogeny. We thank M. D. Ezcurra, G. Mayr, and D. Pol for the photographs of several derived maniraptoran taxa. For financial support we thank CONICET, Agencia Nacional de Promoción Científica y Técnica, and National Geographic Society.

The authors are also indebted to Dr. Jorge Rabassa (CONICET, CADIC), from the Editorial Board of the South America and Southern Hemisphere Sub-series of the Springer Brief Monographies Series on Earth Sciences, who kindly encouraged us to complete this contribution.

Contents

Chapter 1
Introduction

As Chiappe (2009) pointed out, "Deciphering the origin of birds, namely, identifying the closest relatives to the most recent common ancestor of *Archaeopteryx* and modern birds, has been a matter of scientific debate and scrutiny throughout the history of evolutionary biology". Although consensus exists among authors that birds evolved from coelurosaurian theropods, paleontologists still debate about the identification of the group of coelurosaurians that most closely approaches the common ancestor of birds.

Deinonychosauria, a clade of sickle-clawed predatory dinosaurs including the families Troodontidae and Dromaeosauridae (Gauthier 1986), has been usually considered as the sister group of birds, and consequently are of prime importance to understand avian origins and their early evolution. Sister group relationships between birds and deinonychosaurians have been reported by most authors (e.g., Gauthier 1986; Forster et al. 1998; Rauhut 2003; Turner et al. 2007a, b; Senter et al. 2004; Senter 2007; Xu et al. 2003; Hwang et al. 2002; Makovicky et al. 2005; Novas and Pol 2005; Novas et al. 2009; Xu et al. 2008, 2011a; see Fig. 1.1), and the name Paraves was coined for the group that joins Deinonychosauria and Aves (Sereno 1997).

One of the best known examples of deinonychosaurian coelurosaurs is *Deinonychus antirrhopus* (Ostrom 1969), which constituted for long time the principal source of anatomical similarities with the early bird *Archaeopteryx* (Ostrom 1976). However, the last 20 years witnessed the discovery of a wide array of avian-like theropods that has considerably amplified the anatomical disparity among deinonychosaurians, some of which resembling more to *Archaeopteryx* rather than to *Deinonychus* (e.g., Xu et al. 1999, 2000, 2003; Norell et al. 2001; Makovicky et al. 2005; Turner et al. 2007a, b; Senter 2007; Hu et al. 2009; Zheng et al. 2009; Novas et al. 2009). Following this, a recent paper of Xu et al. (2011a) indicate that *Archaeopteryx* was probably more nearly related to deinonychosaurians rather than to birds (see also Paul 2002; see Fig. 1.1). Among these newly discovered theropods that show remarkable bird-like characteristics are the four-winged theropods *Microraptor* (Xu et al. 2000, 2003) and *Anchiornis* (Xu et al. 2008; Hu et al. 2009), and the unenlagiids *Unenlagia, Buitreraptor,* and *Rahonavis*. Besides, a bizarre group of minute-sized coelurosaurs, the Scansoriopterygidae, also exhibits some avian similarities that lead some authors (i.e., Zhang et al. 2008; Hu et al. 2009) to interpret them as more closely related to birds than other dinosaurs.

F. L. Agnolín and F. E. Novas, *Avian Ancestors*, SpringerBriefs in Earth System Sciences, DOI: 10.1007/978-94-007-5637-3_1, © The Author(s) 2013

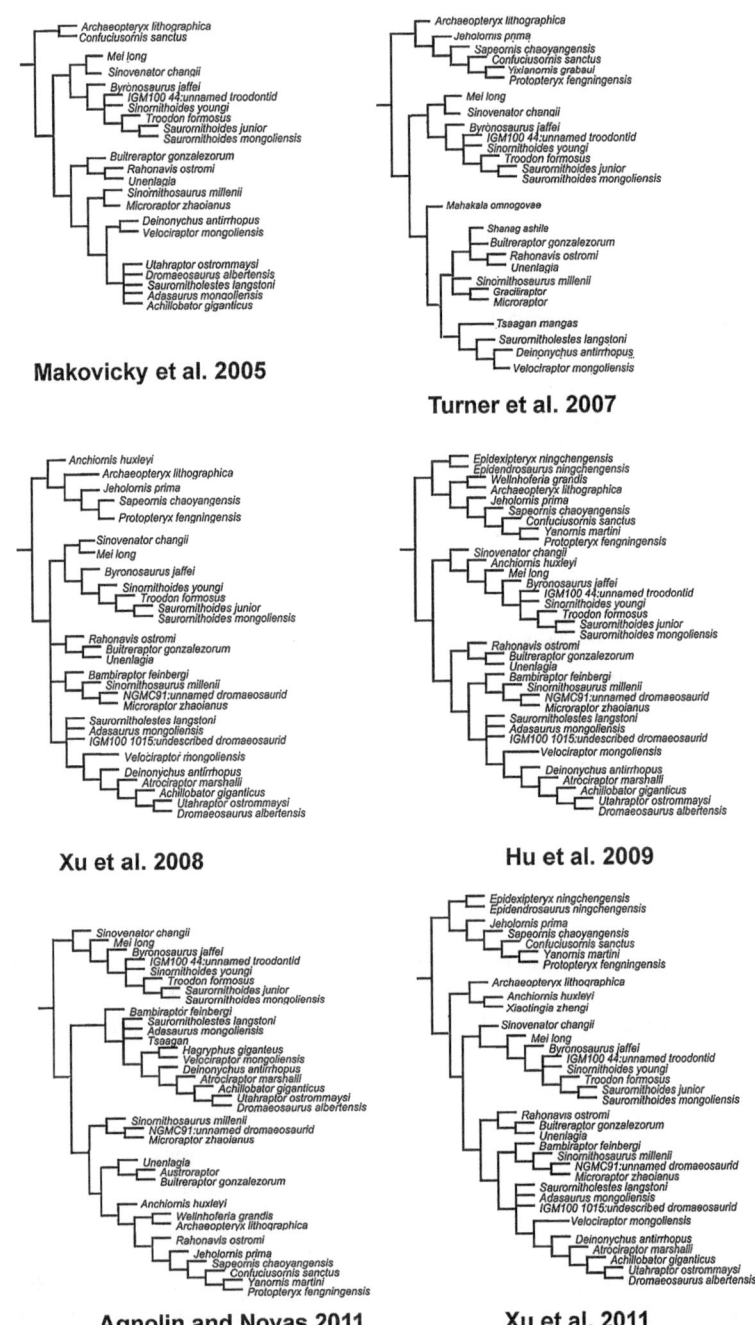

Fig. 1.1 Diagrams showing phylogenetic hypotheses for Paraves advocated by six previous phylogenetic analyses

Here we present evidence that Microraptoria, Unenlagiidae, and *Anchiornis* form successive sister taxa of Aves, and that the Scansoriopterygidae are basal coelurosaurians not directly related to birds. The implications in the evolutionary sequence of anatomical characters leading to birds, including the origin of flight, are also considered in light of this new phylogenetic hypothesis.

References

Chiappe LM (2009) Downsized dinosaurs: The evolutionary transition to modern birds. Evo Edu Outreach 2:248–256.

Forster C, Sampson S, Chiappe LM, Krause D (1998) The theropod ancestry of birds: new evidence from the Late Cretaceous of Madagascar. Science 279:1915–1919

Gauthier JA (1986) Saurischian monophyly and the origin of birds. Mem Calif Acad Sci 8:1–46

Hu D, Hou L, Zhang L, Xu X (2009) A pre-*Archaeopteryx* troodontid theropod from China with long feathers on the metatarsus. Nature 461:640–643

Hwang SH, Norell MA, Qiang J, Keqin G (2002) New specimens of *Microraptor zhaoianus* (Theropoda: Dromaeosauridae) from Northeastern China. Amer Mus Novit 3381:1–44

Makovicky PJ, Apesteguía S, Agnolín FL (2005) The earliest dromaeosaurid theropod from South America. Nature 437:1007–1011

Norell MA, Clark JM, Makovicky PJ (2001) Phylogenetic relationships among coelurosaurian theropods. In: Gall LF, Gauthier J (eds) New perspectives on the origin and early evolution of birds. Peabody Museum of Natural History, Yale University, New Haven, pp 49–68

Novas FE, Pol D (2005) New evidence on deinonychosaurian dinosaurs from the Late Cretaceous of Patagonia. Nature 433:858–861

Novas FE, Pol D, Canale JI, Porfiri JD, Calvo JO (2009) A bizarre Cretaceous theropod dinosaur from Patagonia and the evolution of Gondwanan dromaeosaurids. Proc Royal Soc London B 126:1101–1107

Ostrom JH (1969) Osteology of *Deinonychus antirrhopus*, an unusual theropod from the Lower Cretaceous of Montana. Bull Peabody Mus Nat Hist 30:1–165

Ostrom JH (1976) *Archaeopteryx* and the origin of birds. Biol J Linn Soc 8:91–182

Paul GS (2002) Dinosaurs of the air. The John Hopkins University Press, Baltimore

Rauhut OWM (2003) The interrelationships and evolution of basal theropod dinosaurs. Spec Pap Palaeont 69:1–213

Senter P (2007) A new look at the phylogeny of Coelurosauria (Dinosauria: Theropoda). J Syst Palaeont 5:429–463

Senter P, Barsbold R, Britt B, Burnham D (2004) Systematics and evolution of Dromaeosauridae (Dinosauria: Theropoda). Bull Gunma Mus Nat Hist 8:1–20

Sereno PC (1997) The origin and evolution of dinosaurs. Ann Rev Earth Planet Sci 25:435–489

Turner AH, Hwang SH, Norell MA (2007a) A small derived theropod from Oösh, Early Cretaceous, Baykhangor Mongolia. Amer Mus Novit 3557:1–27

Turner AH, Pol D, Clarke JA, Erickson GM, Norell MA (2007b) A basal dromaeosaurid and size evolution preceding avian flight. Science 317:1378–1381

Xu X, Wang X-L, Wu X-C (1999) A dromaeosaurid dinosaur with a filamentous integument from the Yixian Formation of China. Nature 401:262–266

Xu X, Zhou Z, Wang X (2000) The smallest known non-avian theropod dinosaur. Nature 408:705–708

Xu X, Zhou Z, Wang X, Huang X, Zhang F, Du X (2003) Four winged dinosaurs from China. Nature 421:335–340

Xu X, Zhao Q, Norell MA, Sullivan C, Hone D, Erickson PG, Wang X, Han F, Guo Y (2008) A new feathered dinosaur fossil that fills a morphological gap in avian origin. Chin Sci Bull 54:430–435

Xu X, You H, Du K, Han F (2011a) An *Archaeopteryx*-like theropod from China and the origin of Avialae. Nature 475:465–470

Zhang F, Zhou Z, Xu X, Wang X, Sullivan C (2008) A bizarre Jurassic maniraptoran from China with elongate ribbon-like feathers. Nature 455:1105–1108

Zheng X, Xu X, You H, Zhao Q, Dong Z (2009) A short-armed dromaeosaurid from the Jehol Group of China with implications for early dromaeosaurid evolution. Proc Royal Soc London B 277:211–217

Chapter 2
Materials and Methods

Institutional Abbreviations

HMN	Museum für Naturkunde, Berlin, Germany;
IVPP V	The Institute of Vertebrate Paleontology and Paleoanthropology, Beijin, China;
MCF PVPH	Museo Carmen Funes, Plaza Huincul, Neuquén, Argentina;
MML	Museo Municipal de Lamarque, Lamarque, Río Negro, Argentina;
MPCA	Museo Provincial Carlos Ameghino, Cippolletti, Río Negro, Argentina;
UA	University of Antananarivo, Madagascar;
YPM	Yale Peabody Museum, Yale, USA.

In order to study sequences of appearance of synapomorphies in the theropod line to birds, we have polarized ingroup characters in the data matrix based on outgroup comparisons (i.e., *Allosaurus* and *Sinraptor*). Most characters are coded as binary (0, plesiomorphic; 1, apomorphic), with the exception of some multistate characters, for which 0 is plesiomorphic and 1,2, and/or 3 represents apomorphic states, considered as a progressive sequence. Question mark (?) indicates that the character state is unknown in available specimens. The script (-) indicates that due to the high apomorphic modifications of the taxon, the character state cannot be checked. These codifications are tabulated in a data matrix to show the distributions of character states.

With the aim to analyze the phylogenetic relationships of Dromaeosauridae with respect to other paravians we performed a phylogenetic analysis using the most recent version of the data matrix published by the TwiG, presented by Hu et al. (2009) and modified by Agnolín and Novas (2011). Definitions of characters 1 through 363 follow Hu et al. (2009); characters 364 through 366 have been modified from Novas et al. (2009). We added characters 369–412 from Xu et al. (2008) dataset. Character 367 is from Gianechini et al. (2009), character 368 has been added from Zheng et al. (2009), characters 413–415 have been added from Xu (2002), characters 416–426 were added from Xu et al. (2011a), and characters 427–429 are added from Osmólska et al. (2004). We have modified character 240 from Hu et al. (2009), in reducing the number of character states from 4 to 2. In this way, the present analysis contains one of the most comprehensive dataset employed up to now, consisting of 88 taxa scored

F. L. Agnolín and F. E. Novas, *Avian Ancestors,* SpringerBriefs in Earth System Sciences, DOI: 10.1007/978-94-007-5637-3_2, © The Author(s) 2013

Fig. 2.1 Phylogenetic
analysis of derived
coelurosaurian theropods.
Present strict consensus
tree depicts Microraptoria,
Unenlagiidae, and
(*Anchiornis* + *Xiaotingia*)
as succesive sister-groups
of Avialae. *Abbreviations*
Drom., Dromaeosauridae

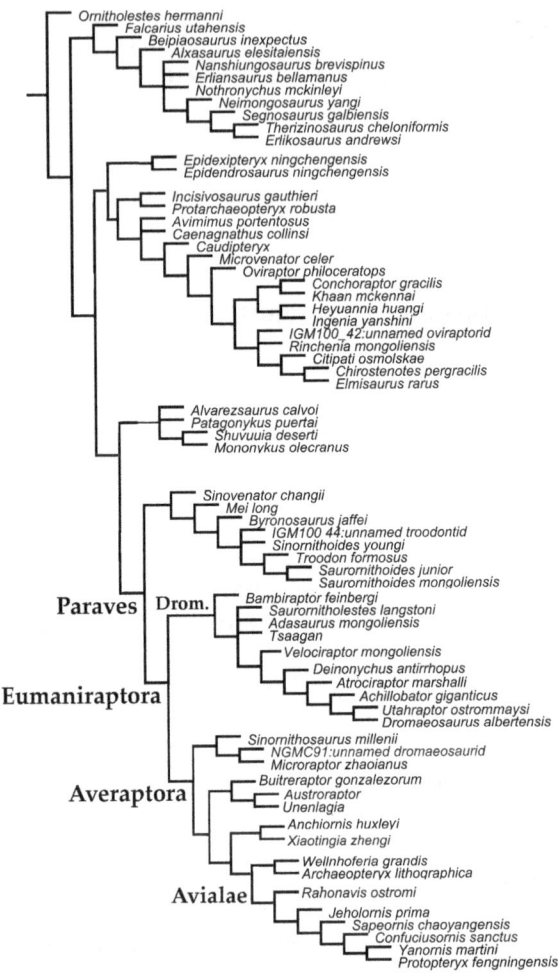

for 429 characters (see Appendix 1, 2). Codifications by previous authors were analyzed in detail, and consequently, several characters from the combined data matrix
of Xu et al. (2008), Hu et al. (2009) and Agnolín and Novas (2011), and Xu et al.
(2011a) were re-scored (Appendix 2).

Phylogenetic analysis was performed using TNT 1.1 (Goloboff et al. 2008). All
characters were equally weighted and treated as unordered. A heuristic search was
performed with 10,000 replicates of the tree bisection and reconnection (TBR)
branch-swapping algorithm. The maximum number of trees saved for each ran-
dom addition sequence replicate was set to 100.

The phylogenetic analysis resulted in the recovery of 50 Most Parsimonious
Trees (MPTs), which bring a Strict Consensus Tree of 1841 steps, with a
consistency index of 0.27, and a retention index of 0.69 (Fig. 2.1).

As proposed by Makovicky et al. (2005), we consider *Neuquenraptor argentinus* Novas and Pol (2005) as junior synonym of *Unenlagia comahuensis* Novas and Puerta 1997 (contra Porfiri et al. 2011), and we follow Agnolín and Novas (2011) in the use of the original family name Unenlagiidae Bonaparte (1999) (instead of Unenlagiinae; sensu Makovicky et al. 2005), to emphasize the distinctiveness of this theropod group.

References

Agnolín FL, Novas FE (2011) Unenlagiid theropods: are they members of Dromaeosauridae (Theropoda, Maniraptora). An Acad Bras Ciênc 83:117–162

Bonaparte JF (1999) Tetrapod faunas from South America and India: a palaeobiogeographic interpretation. Proc India Acad Sci 65:427–437

Gianechini FA, Apesteguía S, Makovicky PJ (2009) The unusual dentition of *Buitreraptor gonzalezorum* (Theropoda, Dromaeosauridae), from Patagonia, Argentina: new insights on the unenlagine teeth. Ameghiniana 52:36A

Goloboff PA, Farris JS, Nixon KC (2008) TNT, a free program for phylogenetic analysis. Cladistics 24:774–786

Hu D, Hou L, Zhang L, Xu X (2009) A pre-*Archaeopteryx* troodontid theropod from China with long feathers on the metatarsus. Nature 461:640–643

Makovicky PJ, Apesteguía S, Agnolín FL (2005) The earliest dromaeosaurid theropod from South America. Nature 437:1007–1011

Novas FE, Pol D (2005) New evidence on deinonychosaurian dinosaurs from the Late Cretaceous of Patagonia. Nature 433:858–861

Novas FE, Puerta P (1997) New evidence concerning avian origins from the Late Cretaceous of NW Patagonia. Nature 387:390–392

Novas FE, Pol D, Canale JI, Porfiri JD, Calvo JO (2009) A bizarre cretaceous theropod dinosaur from Patagonia and the evolution of Gondwanan dromaeosaurids. Proc Royal Soc London B 126:1101–1107

Osmólska H, Currie PJ, Barsbold R (2004) Oviraptorosauria. In: Weishampel DB, Dodson P, Osmolska H (eds) The Dinosauria, 2nd edn. University of California Press, Berkeley, pp 165–183

Porfiri JD, Calvo JO, Santos D (2011) A new small deinonychosaur (Dinosauria: Theropoda) from the Late Cretaceous of Patagonia, Argentina. An Acad Bras Ciênc 83:109–116

Xu X (2002) Deinonychosaurian fossils from the Jehol Group of western Liaoning and the coelurosaurian evolution. Dissertation for the doctoral degree. Chinese Academy of Sciences, Beijing

Xu X, Zhao Q, Norell MA, Sullivan C, Hone D, Erickson PG, Wang X, Han F, Guo Y (2008) A new feathered dinosaur fossil that fills a morphological gap in avian origin. Chinese Sci Bull 54:430–435

Xu X, You H, Du K, Han F (2011a) An *Archaeopteryx*-like theropod from China and the origin of Avialae. Nature 475:465–470

Zheng X, Xu X, You H, Zhao Q, Dong Z (2009) A short-armed dromaeosaurid from the Jehol Group of China with implications for early dromaeosaurid evolution. Proc Royal Soc London B 277:211–217

Chapter 3
Systematic Palaeontology

3.1 Paraves Sereno, 1997

Definition. All maniraptorans are closer to Aves (= Neornithes) than to *Oviraptor* (Sereno 1998).

Synapomorphies. 71(1), 103(1), 175(2), **180(1)**, 296(1), 297 (1), 300(1), 319(1), **330(2)**, 342(1), 374(1), 377(1).

Comments. Several characters here considered as diagnostic of Paraves (*sensu* Gauthier's 1986), were previously proposed as diagnostic of Dromaeosauridae, including 71(1), 103(1), 175(2), 296(1) (Norell and Makovicky 2004; Makovicky et al. 2005; Zheng et al. 2009), and are here recovered as diagnostic of a more inclusive node (i.e. Paraves; see Agnolín and Novas, 2011).

In the first cladistic phylogenetic analysis of Saurischia conducted by Gauthier (1986), the term Deinonychosauria Colbert and Russell 1969, was employed to include the theropod groups Troodontidae and Dromaeosauridae. The monophyly of Deinonychosauria was accepted by most latter authors (see Norell et al. 2001; Xu 2002; Makovicky et al. 2005). Sereno's (1998) defined Deinonychosauria as *Troodon*, *Velociraptor*, their most common ancestor and all descendants. If we apply Sereno's definition to the present phylogeny, Deinonychosauria becomes a senior synonym of Paraves. However, for the sake of clarity, we prefer to use here the term Paraves for the clade including theropods more derived than Troodontidae, and consider Deinonychosauria as a paraphyletic group, in contrast with most previous proposals.

3.2 Eumaniraptora Padian, Hutchinson, and Holtz, 1999

Definition. The theropod group that includes all taxa closer to *Passer* than to *Troodon*.

Synapomorphies. 54(1) lateral border of quadrate shaft with lateral tab that touches squamosal and quadratojugal above an enlarged quadrate foramen; 71(1) dentary with subparallel dorsal and ventral edges; 85(0) dentary and maxillary teeth large, less than 25 in dentary; 103(1) parapophyses of posterior trunk vertebrae

F. L. Agnolín and F. E. Novas, *Avian Ancestors*, SpringerBriefs in Earth System Sciences, DOI: 10.1007/978-94-007-5637-3_3, © The Author(s) 2013

distinctly projected on pedicels; 125(1) ossified uncinate processes present; 175(2) pubis moderately posteriorly oriented; **180(1)** femur with circular fovea present in center of medial surface of head; 296(1) dorsal surface of manual ungual I arches higher than level of dorsal extremity of proximal articular surface; 297(1) dorsal surface of manual ungual II arches higher than level of dorsal extremity of proximal articular surface; 300(1) proximodorsal 'lip' on manual unguals II and III present; 319(1) hallucal ungual strongly curved; **330(2)** lateral face of ischial shaft with longitudinal ridge dividing lateral surface into anterior and posterior parts; 342(1) pedal phalanx II-1 shorter than pedal phalanx IV-1; 374(1) low crest on the lateral surface of the scapula continuous from the dorsal margin of the acromion process; 377(1) scapula relatively robust, only sharply ridged along the dorsal margin close to the distal end (Fig. 3.1).

Content. This clade includes Dromaeosauridae + (Microraptoria + (Unenlagii dae ((*Anchiornis* + *Xiaotingia*) + Avialae))).

Comments. In his detailed analyses, Ostrom (1969, 1976) proposed Dromaeosauridae as the non-avian theropod group more nearly related to birds. However, since Gauthier's paper (1986), most recent authors consider

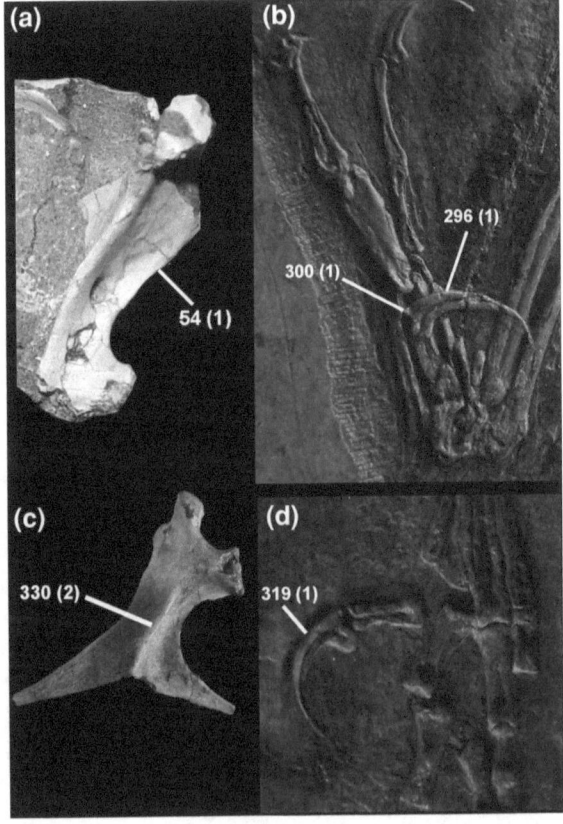

Fig. 3.1 Selected synapomorphies of Eumaniraptora. **a** right quadrate of *Buitreraptor gonzalezorum* (MPCA 245, holotype) in posterior view, showing the presence of a lateral tab (ch. 54-1). **b** left hand of *Confuciusornis sanctus* (IVPP V11374-5) showing a strongly dorsally arched manual ungual II (ch. 297-1), and a proximodorsal lip (ch. 300-1). **c** right ischium of *Buitreraptor gonzalezorum* (MPCA 245, holotype) in lateral view, showing the presence of a longitudinal ridge (ch. 330-2). **d** right foot of *Confuciusornis sanctus* (IVPP V11374-5) in posterior view, showing strongly arched hallucal ungual (ch. 319-1). Not to scale

Dromaeosauridae and Troodontidae as sister groups, both conforming to a monophyletic Deinonychosauria. Deinonychosaurs as a whole are considered by most authors as the sister group of Avialae (Norell et al. 2001; Xu 2002; Xu and Norell 2004; Xu and Wang 2004; Xu et al. 2000, 2008; Makovicky et al. 2005; Senter 2007; Turner et al. 2007a, b; Zheng et al. 2009; Novas et al. 2009; Agnolín and Novas 2011). However, some authors have doubt on deinonychosaurian monophyly. Currie (1985, 1987, 2000) proposed that deinonychosaurs were a non-monophyletic group and noted several similitudes between troodontids and birds. On the other hand, Senter et al. (2004a, b) proposed a non-monophyletic Deinonychosauria, and concluded that Dromaeosauridae was the sister group of Avialae. More recently, Agnolín and Novas (2011) analyzed in detail deinonychosaurian synapomorphies, concluding that several putative-derived characters were more widespread than previously thought. In this paper Deinonychosauria is considered as a paraphyletic assemblage, and Dromaeosauridae is recovered as more nearly related to birds than to troodontids, as suggested earlier by Ostrom (1969, 1976) and Senter et al. 2004a, b (Fig. 2.2). 15 derived features support Dromaeosauridae as more derived than troodontids, and the clade including dromeaosaurids and birds is here named Eumaniraptora, as originally defined by Padian et al. (1999).

3.3 Averaptora new clade

Definition. The theropod group that includes all taxa closer to *Passer* than to *Dromaeosaurus*.

Synapomorphies. 90(1) interdental plates located medially between teeth; 162(0) antitrochanter posterior to acetabulum reduced or absent; 200(1) metatarsal III pinched proximally; 290(1) presence of a posterior-medial flange on manual phalanx II-1; 295(1) manual phalanx I-1 bowed (with its palmar surface concave); 320(2) length of pedal phalanx II-2 $\geq 1 \times$ length of phalanx II-1 (Fig. 3.2).

Content. This node includes the clades Microraptoria + (Unenlagiidae ((*Anchiornis* + *Xiaotingia*) + Avialae)).

Comments. The clade Microraptoria was coined by Senter et al. (2004a, b) to include *Microraptor* and other Cretaceous forms from China and North America, such as *Sinornithosaurus* and *Bambiraptor*. More recently, other taxa have also been recognized as members of this group (i.e., *Graciliraptor*, *Hesperonychus*, and *Tianyuraptor*; Xu and Wang 2004; Longrich and Currie 2008; Zheng et al. 2009). It must be mentioned that *Bambiraptor* was reinterpreted by Longrich and Currie (2008) as a member of Saurornitholestinae, and that features supporting *Tianyuraptor* within Microraptoria are debatable.

Microraptor comes from the Jiufotang Formation (Early Cretaceous) of NE China (Xu et al. 2000, 2003), and is currently represented by the species *M. zhaoianus* (Xu et al. 2000) and *M. gui* (Xu et al. 2003). Several specimens of *M.gui* preserve a feathered covering, demonstrating that elongate pennaceous feathers

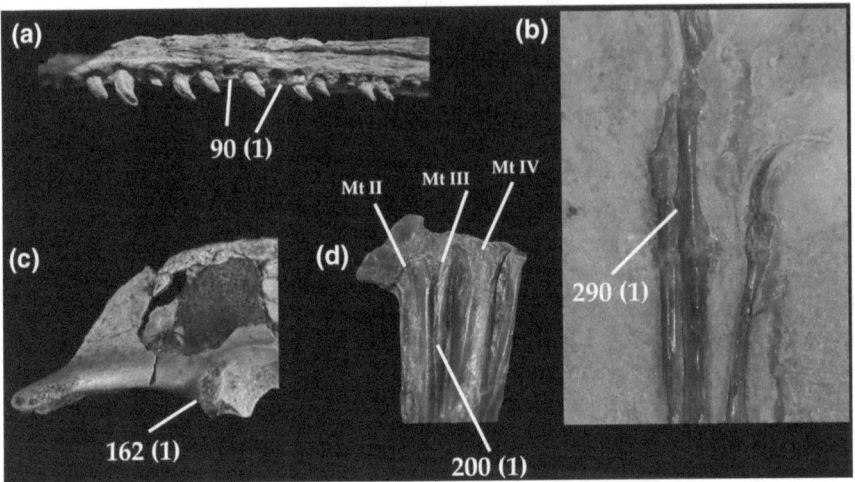

Fig. 3.2 Selected synapomorphies of Averaptora. **a** left dentary of *Austroraptor cabazai* (MML 195, holotype) in dorsal view, showing the presence of interdental plates (ch. 90-1). **b** left hand of *Archaeopteryx lithographica* (HMN 1880) showing a posterolateral flange on phalanx II-1 (ch. 290-1). **c** posterior end of right ilium of *Buitreraptor gonzalezorum* (MPCA 245, holotype) in lateral view, showing a reduced antitrochanter (ch. 162-1). **d** right metatarsus in posterior view of *Sinornithosaurus milleni* (IVPP V12811; holotype), showing proximally pinched metatarsal III (ch. 200-1). Not to scale. *Abbreviations* metatarsal (*Mt*)

were present on both fore- and hindlimbs (Xu et al. 2003), thus resulting in a four-winged condition currently unknown in both extant and extinct flying birds. Such peculiar four-winged patterns heated the debate about the origin of birds and their flight, leading some authorities (e.g., Padian and Ricqlès 2009) to discredit *Microraptor* as a source of information about the acquisition of features in the line to birds.

Microraptorians have been regarded by most authors (e.g., Norell et al. 2001; Xu et al. 2000, 2003; Hwang et al. 2002; Senter et al.2004a; Senter 2007; Novas et al. 2009; Hu et al. 2009) as the sister group of the remaining dromaeosaurids (i.e., Eudromaeosauria sensu Longrich and Currie 2008). However, we have demonstrated elsewhere (Agnolín and Novas 2011) that the characters supporting this view (e.g., stalk-like trunk parapophyses; size difference between anterior and posterior denticles on maxillary teeth; tooth root unconstricted; metatarsal II with distal ginglymoid articulation) are widespread among basal paravians.

Skeletal information supporting the basal position of *Microraptor* among dromaeosaurids, taken together with the peculiar four-winged condition, has led some authors (Senter et al. 2004a, b; Senter 2007) to hypothesize that aerial locomotion and arboreality were acquired independently in microraptorians and ornithuran birds. However, this analysis depicts microraptorians outside Dromaeosauridae and Deinonychosauria, and most of the avian-like features are interpreted here as synapomorphies uniting them with unenlagiids, *Anchiornis*, and avialans. In our view, aerial locomotion evolved once, in the common ancestor of microraptorians and birds.

In the present analysis we recover Microraptoria, Unenlagiidae, and *Anchiornis* as successive sister groups of birds. The clade Microraptoria was usually considered as included within Dromaeosauridae. Diagnostic features cited by previous authors include: 36-1, quadratojugal Y- or T-shaped (character 90-1 of Xu et al. 1999); 40-1, lacrimal "T"-shaped (character 86-1 of Xu et al. 1999); 43-1, anterior margin of supratemporal fossa sinusoidal (character 42-1 of Makovicky et al. 2005); 46-1, supratemporal fossa covering most frontal processes of postorbital and extending anteriorly on dorsal surface of frontal to at least the level of posterior orbital margin (character 88-1 of Xu et al. 1999); 54-1, quadratojugal fenestra widely opened (character 91-1 of Xu et al. 1999); 54-1, quadrate shaft with a lateral process (character 53-1 of Makovicky et al. 2005); 71-1, upper and ventral margins of dentary subparallel (character 96-1 of Xu et al. 1999); 88-1, tooth root unconstricted (character 88-1 of Makovicky et al. 2005); 89-1, size difference between between anterior and posterior denticles on maxillary teeth (Xu and Wang 2004); 95-0, distally positioned cervical epipophyses (character 95-1 of Makovicky et al. 2005); 103-1, trunk parapophyses stalk-like (character 103-1 of Makovicky et al. 2005); 122-1, ossified caudal rods extending lengths of prezygapophyses and chevrons (character 100-1; Xu et al. 1999); 123-2, chevrons bifurcate at both ends (character 123-2 of Makovicky et al. 2005); 198-1, metatarsal II with ginglymoid articulation (character 201-1 of Makovicky et al. 2005); 249-1, manual phalanx III-2 significantly shortened (Xu and Wang 2004); 296-1, dorsal margin of manual unguals arches high over dorsal extremity of proximal articular facets (characters 60-1 and 69-2 of Senter et al. 2004a, b); 315-1, long metatarsal V (Xu et al. 2003) (Figs. 3.3, 3.4). Of the above-mentioned features, characters 36-1, 40-1, 43-1, 54-1, 71-1, 88-1, 89-1, 95-0, 103-1, 123-1, 198-1, and 296-1 were dismissed as dromaeosaurid synapomorphies (Agnolín and Novas 2011) because they are widely distributed among paravian theropods. Moreover, stalk-like projected parapophyses on dorsal vertebrae (character 103-1) were recently described for the troodontids *Mei* and *Talos* (Zanno et al. 2011). Regarding character 122-1, it may be considered as a probable synapomorphy of Microraptoria + Dromaeosauridae; however, it was here recovered as diagnostic of Dromaeosauridae + Averaptora. With respect to character 249-1, the derived condition of a shortened manual phalanx III-2 is present at least in *Archaeopteryx* and the enantiornithine *Protopteryx* (Zhang and Zhou 2000; Paul 2002), but are still elongate in *Jeholornis* and *Confuciusornis* (Chiappe et al. 1999; Zhou and Zhang 2002), suggesting a complex and probably more widespread condition of this character. In regard to character 315-1, the elongation of metatarsal V was not quantified by previous authors, so this character appears to be ambiguous. Moreover, the metatarsal V remains unpreserved in any of the known Unenlagiidae. In sum, we do not find evidence supporting Microraptoria as members of Dromaeosauridae.

Fig. 3.3 *Sinornithosaurus milenii* (Holotype, IVPP V12811). **a-b** skull, **a** counterslab, **b** slab. *Abbreviations* angular (*ang*), dentary (*dent*), feathers (*fe*), frontal (*fr*), lacrimal (*lac*), parietal (*par*), postorbital (*po*), quadrate ramus of pterigoid (*qrp*), surangular (*sa*), splenial (*spl*). Scale bar 3 cm

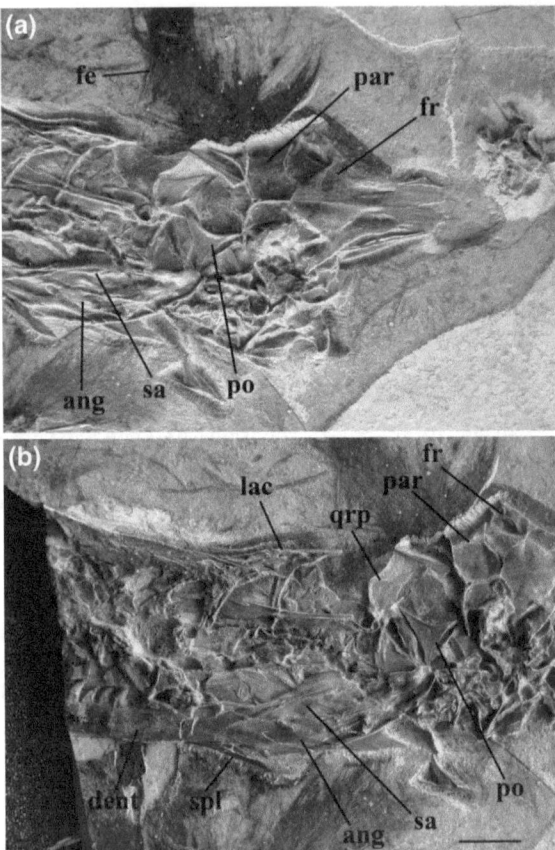

On the other hand, Turner et al. (2007a) considered Microraptoria to be the sister group of Unenlagiidae, and both taxa were nested by these authors within Dromaeosauridae. They reported seven synapomorphies uniting microraptorans and unenlagiids (characters 123-2, 139-1, 143-1, 203-1, 229-1, 234-1, 244-1, and 333-1 in Turner et al. 2007a). Characters 123-2 (chevrons bifurcate at both ends), 203-1 (subarctometatarsalian metatarsus), and 333-1 (longitudinal flange along caudal surface of metatarsal IV) were discussed in detail by Agnolín and Novas (2011), and are here considered as more widespread than previously thought. Another feature recognized by Turner et al. (2007a) to unite Unenlagiidae and Microraptoria is humerus longer than the scapula (character 139-1) which is present in *Anchiornis*, *Xiaotingia*, *Archaeopteryx*, *Jeholornis*, and more derived birds (Zhou and Zhang 2002; Mayr et al. 2007; Hu et al. 2009; Xu et al. 2011a), and thus it is probably diagnostic of a more inclusive clade than Unenlagiidae + Microraptoria clade. Another condition cited as diagnostic of Unenlagiidae + Microraptoria is the presence of a convex distal articular surface of ulna (character 143-1), but this condition appears to be more widespread

Fig. 3.4 *Sinornithosaurus milenii* (Holotype, IVPP V12811). **a** caudal vertebrae, **b** right foot in posterior view. *Abbreviations* astragalus (*astr*), haemal arch (*ha*), metatarsal (*mt*), rod-like process of prezygapophyses (*rp*), vertebral centrum (*vc*). Scale bar 2 cm

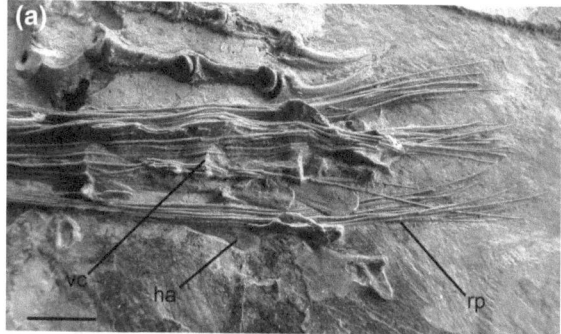

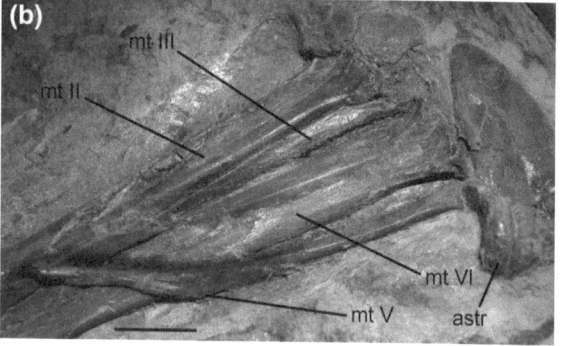

than previously suggested, being also documented in *Anchiornis*, *Xiaotingia*, *Archaeopteryx*, *Rahonavis*, and *Confuciusornis* (Forster et al. 1998; Chiappe et al. 1999; Mayr et al. 2007; Campbell 2008; Xu et al. 2008, 2011a). The presence of an obturator process of ischium strongly extended rostrally, and showing a short base (character 234-1 of Turner et al. 2007a) also exhibits an ambiguous phylogenetic significance. In fact, the distal end of the ischium is strongly variable in unenlagiids (e.g. *Unenlagia*, *Buitreraptor*; Novas and Puerta 1997; Makovicky et al. 2005), microraptorians (e.g. *Microraptor*, *Sinornithosaurus*; Xu 2002), *Anchiornis* (Xu et al. 2008; Hu et al. 2009), and basal avialans (e.g. *Rahonavis*, *Archaeopteryx*; Forster et al. 1998; Mayr et al. 2007) (Fig. 3.5). However, in *Microraptor*, *Anchiornis*, and *Rahonavis* the distal end of the obturator process of ischium is similar in being thin and long and in showing its distal margin distally concave (Forster et al. 1998; Xu et al. 2008). On the other hand, the ischium of *Sinornithosaurus* is similar to that of *Buitreraptor*, *Unenlagia*, and dromaeosaurids in having a proximally located obturator process, whereas in *Microraptor*, *Anchiornis*, and Avialae this process is continuous with the distal margin of the ischium, being subhorizontally oriented (Xu 2002; Hu et al. 2009) (Fig. 3.5). Moreover, in *Archaeopteryx* and *Anchiornis* the distal margin of the ischium is strongly concave and exhibits an additional distal posterodorsal process, a trait not seen in any other known theropod. Finally, character 244-1 of Turner et al. (2007a), defined as "presence of the lateral lamina of the anteroventral

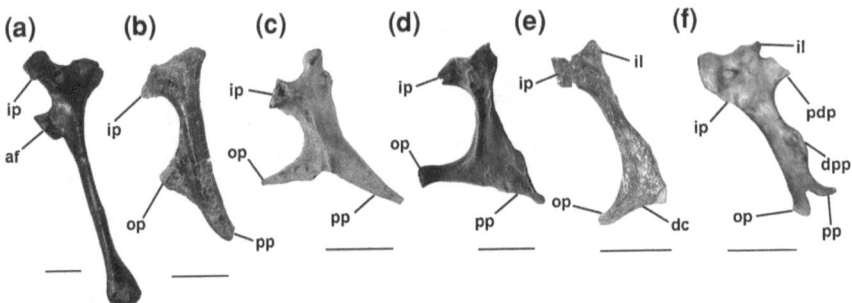

Fig. 3.5 Left ischia of selected theropod taxa in left lateral view. **a** *Allosaurus fragilis* (YPM 14554), **b** *Deinonychus antirrhopus* (YPM 5235), **c** *Buitreraptor gonzalezorum* (MPCA 245, holotype), **d** *Sinornithosaurus milenii* (Holotype, IVPP V12811), **e** *Microraptor gui* (IVPP V 13352, holotype), **f** *Archaeopteryx lithographica* (BMNH 37001). *Abbreviations* anterior flange (*af*), distal concavity (*dc*), additional dorsal posterior process (*dpp*), ischiatic process for articulation with pubis (*ip*), obturator process (*op*), posterodorsal process (*pdp*), posterior process (*pp*), Scale bar **a** 8 cm, **b** 4 cm, **c, d** 2 cm, **e, f** 1 cm

ramus of the nasal small with a narrow and triangular exposure", is absent in Unenlagiidae and Microraptoria. However, this trait is present in a large variety of dromaeosaurids and troodontids (e.g. *Byronosaurus, Sinusonasus, Linhevenator, Bambiraptor, Velociraptor*; Barsbold and Osmólska 1999; Burnham et al. 2000; Norell et al. 2000; Xu and Wang 2004; Xu et al. 2011b). Moreover, this condition is also seen in derived birds, including Enantiornithes (Chiappe and Walker 2002), *Archaeorhynchus* (Zhou and Zhang 2006) and *Hongshangornis* (Zhou and Zhang 2005). This suggests that character 244-1 is more widely present than proposed by Turner et al. (2007a). In conclusion, a clade made by Unenlagiidae + Microraptoria is not supported by current information.

3.4 Unenlagiidae + Avialae clade

Synapomorphies. 76(0) splenial not widely exposed on lateral surface of mandible; 159(0) postacetabular blades of ilia parallel in dorsal view; **270(3)** acromion triangular, with apex pointing away from and subparallel to scapular blade; 391(1) distal end of ulna with anteroposterior flattening present but weak; 400(1) ischial peduncle of pubis short, flush with the lateral surface of the pubic shaft; 405(1) shaft of the ischium with its minimum anteroposterior length more than 20 % total ischial length (Fig. 3.6).

Content. This node includes the clades Unenlagiidae + ((*Anchiornis* + *Xiaotin gia*) + Avialae).

Comments. The family Unenlagiidae of derived Late Cretaceous paravians includes the Patagonian *Unenlagia comahuensis* (Novas and Puerta 1997; Novas 1999, 2004, 2009), *U. paynemili* (Calvo et al. 2004; Gianechini and Apesteguía

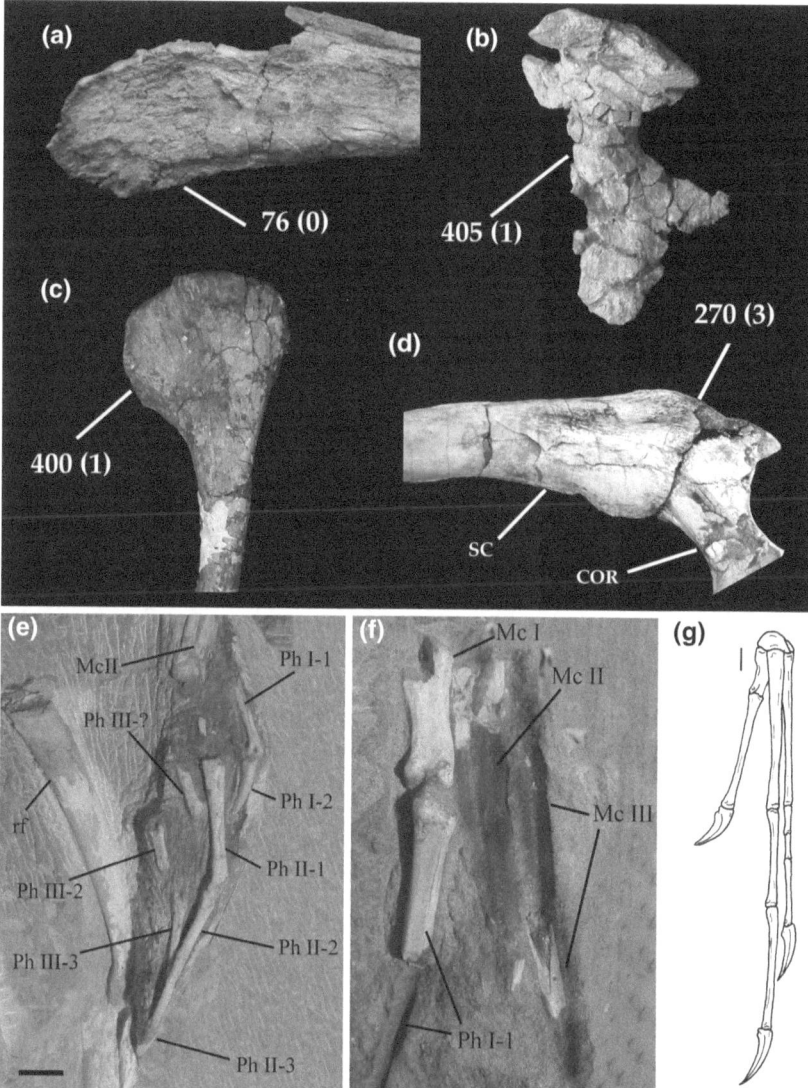

Fig. 3.6 a–d, Selected synapomorphies of Unenlagiidae + Avialae. a right dentary in lateral view of *Austroraptor cabazai* (MML 195, holotype), showing the absence of lateral exposition of splenial bone (ch. 76-0). **b** left ischium of *Unenlagia comahuensis* in lateral view (MCF- PVPH-78), showing anteroposterior width of shaft being more than 20 % of whole length of ischium (ch. 405-1). **c** proximal end of right pubis of *Unenlagia comahuensis* in lateral view (MCF- PVPH-78), showing a short ischiac pedicle (ch. 400-1). **d** right scapulocoracoid of *Buitreraptor gonzalezorum* (MPCA 245, holotype), showing a well-developed and subtriangular acromion (ch. 270-3). **e–g,** *Buitreraptor gonzalezorum*, anatomical details of specimen at the Museo Patagónico de Ciencias Naturales (General Roca, Argentina). **e,** right hand in dorsal view; **f,** incomplete right metacarpals in palmar view, metacarpals II and III preserved as incomplete molds and fragmentary bone; **g,** reconstruction of left hand of *Buitreraptor*. **a–d,** not to scale; **e–g,** scale bar 1 cm. *Abbreviations* metacarpal (*mc*), phalanx (*ph*), *right* femur (*rf*), scapula (*sc*), coracoid(*cor*).

2011), *Neuquenraptor argentinus* (Novas and Pol 2005; Gianechini and Apesteguía 2011), *Buitreraptor gonzalezorum* (Makovicky et al. 2005; Gianechini and Apesteguía 2011), and *Austroraptor cabazai* (Novas et al. 2009; Gianechini and Apesteguía 2011). *Rahonavis ostromi*, from the Upper Cretaceous of Madagascar (Forster et al. 1998) has also been considered by some authors (e.g., Makovicky et al. 2005; Novas and Pol 2005; Novas et al. 2009; Turner et al. 2007a, b) as a member of this clade (but see below). Recently, the genus and species *Pamparaptor micros* was described on the basis of a highly incomplete foot coming from the Upper Cretaceous of Patagonia (Porfiri et al. 2011).

Unenlagia was originally interpreted as more closely related to birds than to dromaeosaurids (e.g. Novas and Puerta 1997; Novas 1999, 2004; Forster et al. 1998; Xu et al. 1999; Rauhut 2003), but Makovicky et al. (2005) considered unenlagiids as nested within Dromaeosauridae, interpretation that found support in most recent cladistic analyses (e.g., Senter 2007; Hu et al. 2009; Novas and Pol 2005; Novas et al. 2009; Xu et al. 2011a). In this phylogenetic context, several derived features of unenlagiids were viewed as examples of evolutionary convergence with birds, and particularly the elongated forelimbs of *Rahonavis* (and its inferred potential for flight capability) were accepted as independently originated from birds (Makovicky et al. 2005; Senter 2007; Novas 2009).

However, in a recent review (Agnolín and Novas 2011) we found that most of the purported dromaeosaurid and deinonychosaurian synapomorphies previously cited for unenlagiines are conflicting, at least, most features are, in fact, more widely distributed among paravians than in dromaeosaurids or deinonychosaurians, instead of others that cannot be identified in unenlagiids due to fragmentary preservation of specimens. The available evidence support, on the contrary, that Unenlagiidae is located as stem Avialae, in agreement with the original proposal (Novas and Puerta 1997).

With regard to particularly *Rahonavis*, it was described as a basal bird by Forster et al. (1998), and although some authors defended its inclusion within Unenlagiidae (e.g., Makovicky et al. 2005; Novas and Pol 2005; Novas et al. 2009), several others (Zhou and Zhang 2002; Hwang et al. 2002; Xu et al. 2008) found the Malagasy taxon allocated within Avialae. Our current analysis also supports *Rahonavis* as a basal bird more derived than *Archaeopteryx*.

Novas and Puerta (1997) in the original description of *Unenlagia* considered that this theropod was more nearly related to *Archaeopteryx* than to Dromaeosauridae. In this way, they considered *Unenlagia* as the basalmost member of the clade Avialae, a criteria that was followed by Agnolín and Novas (2011). However, the name Avialae was originally employed by Gauthier's (1986; see also Gauthier's and De Queiroz 2001) to include *Archaeopteryx* plus Ornithurae birds, and most latter authors used this name for the *Archaeopteryx* node (Perle et al. 1993; Norell et al. 2001; see also Gauthier and De Queiroz 2001). Following those authors we here restrict the name Avialae to the node including *Archaeopteryx* and more derived birds.

The name Unenlagiidae was coined by Bonaparte (1999) to unite *Unenlagia* and *Rahonavis*, on the basis of general similarities of their pelvic anatomy. Later, Makovicky et al. (2005) included Unenlagiidae within Dromaeosauridae, under the name Unenlagiinae, and they defined this clade as "all taxa closer to *Unenlagia*

comahuensis than to *Velociraptor mongoliensis*". However, if we follow the phylogenetic definition used by these authors, under the present phylogeny Unenlagiidae
may be expanded to include several other taxa, such as Microraptoria, *Anchiornis*,
Archaeopteryx, and Ornithurae birds. On the other hand, Turner et al. (2007a) pro-
vided an amended definition of Unenlagiidae (Unenlagiinae therein) and considered
this clade as "all taxa closer to *Unenlagia comahuensis* than to *Velociraptor mongoliensis* and *Microraptor zhaoianus*". Similar to the definition employed by Makovicky
et al. (2005), the definition of Unenlagiidae employed by Turner et al. (2007a,
b, c) may also include a very large array of theropods, as for example *Anchiornis*,
Archaeopteryx, and Ornithurae birds. For the sake of clarity, we prefer to redefine
here Unenlagiidae as follows: "the node including *Unenlagia*, and *Buitreraptor*, its
most common ancestor plus all of its descendants".

Gauthier (1986) indicated that the presence of a prominent and subtriangular
acromion process on scapula for ligamentous connection to the clavicle was diagnostic of Avialae. In fact, the pointed and subtriangular acromion is clearly seen
in *Archaeopteryx* and derived birds, being rather different from the plate-like condition seen in more basal theropods, including *Sinornithosaurus* (Carpenter 2002;
Xu 2002). In fact, the elongate and subtriangular acromion seen in *Buitreraptor*,
Unenlagia, and basal birds (e.g. *Rahonavis*, *Jeholornis*; Forster et al. 1998; Zhou
and Zhang 2002, 2003a,2003b), indicate that the scapula was very probable in near
contact with the clavicles.

Certainly, the most complete and informative unenlagiid is *Buitreraptor gonzalezorum*, which was described on the basis of two individuals representing most of the
skeleton (Makovicky et al. 2005). However, the hands of this taxon (and the remaining unenlagiids) remained nearly unknown, with the single exception of fragmentary
unidentified manual elements recovered with the holotype specimen. A new specimen of *Buitreraptor gonzalezorum* belonging to the Museo Patagónico de Ciencias
Naturales (General Roca, Río Negro, Argentina) preserves fragments of both left
and right forearms, thus allowing a reconstruction of *Buitreraptor* hands (Fig.3.6).
Metacarpal I is relatively gracile and shows subparallel medial and lateral margins;
its distal articulation is slightly medially oriented. Phalanx I-1 is extremely elongate (45 mm) and is subequal in length to phalanx II-2. Digit II is the longest of
the hand, and characterizes for being its gracile proportions, reminiscent to those of
Archaeopteryx (Paul 2002) rather than to the shorter and stouter proportions of non-
avian theropods. Metacarpal II of *Buitreraptor* is relatively robust, but it is poorly
preserved and its morphology cannot be properly known. Phalanx II-1 is very similar
to that of *Archaeopteryx* and dromaeosaurids, showing a transversely expanded proximal end. Phalanx II-2 is elongate and gracile (47 mm) and lacks of a well defined
distal ginglymoid, a condition similar to *Archaeopteryx* and *Confuciusornis* (Paul
2002; Chiappe et al. 1999), different from the distally excavated and well developed
articular surface seen in *Deinonychus* (Ostrom 1969). Phalanx II-2 of *Buitreraptor*
is much more elongate than in remaining theropods, including *Archaeopteryx* (Paul
2002). It lacks the posterolateral flange present in *Sinornithosaurus*, *Archaeopteryx*,
and more derived birds (Paul 2002). The manual ungual of digit II shows its dorsal
arch more dorsally projected that its proximal articulation, a condition also shared

with *Archaeopteryx* and dromaeosaurid theropods (Paul 2002; Zheng et al. 2009). Digit III is incomplete, but the preserved metacarpal and phalanges indicate that they were elongate elements. Phalanx III-3 is very thin and elongate, similar in proportions and morphology to *Archaeopteryx*, *Xiaotingia*, and *Sinornithosaurus* (Paul 2002; Xu et al. 2011). In sum, the hand of *Buitreraptor* approaches basal birds such as *Archaeopteryx* and *Anchiornis* (Paul 2002; Xu et al. 2008) in the extremely elongate and gracile proportions of metacarpals and phalanges, which are much more elongate than in *Microraptor* (Xu 1999) and dromaeosaurids (e.g., *Deinonychus*, *Velociraptor*, Ostrom 1969; Norell and Makovicky 1999). Moreover, digit II is extremely large: the distal end of the ungual phalanx of digit III does not level the distal end of phalanx II-2. In the same way, ungual phalanx of digit I does not approach the distal end of phalanx II-1. The extremely elongate digit II is a feature that *Buitreraptor* shares with *Sinornithosaurus* and birds, including *Archaeopteryx*, *Jeholornis* and *Confuciusornis* (Chiappe et al. 1999; Paul 2002; Zhou and Zhang 2002). However, digits of *Buitreraptor* appear to be even more elongate than in any known theropod (including birds) and may constitute an autapomorphy for this genus (combined length of phalanges of digit II/femur length ratio: 0.67).

3.5 *(Anchiornis + Xiaotingia)* + **Avialae clade**

Synapomorphies. 21(1) internarial bar flat; 262(1) length of mid-cervical centra markedly longer than dorsal centra; 330(0) lateral face of ischial shaft flat (or round in rodlike ischia); 342(0) pedal phalanx II-1 longer than pedal phalanx IV-1;

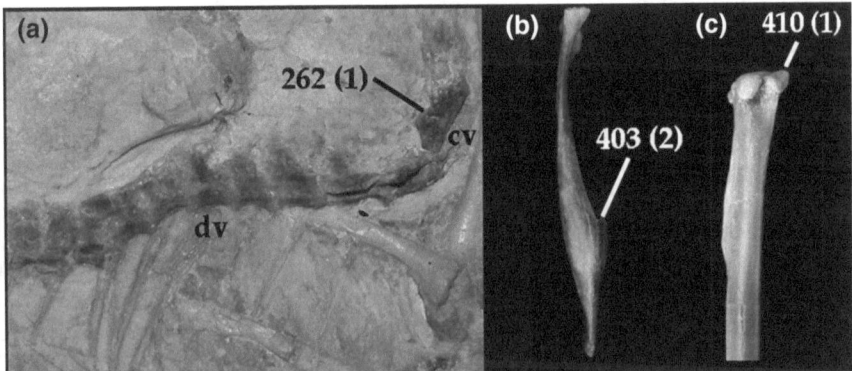

Fig. 3.7 Selected synapomorphies of ((*Anchiornis* + *Xiaotingia*) + Avialae). **a** vertebral column of *Archaeopteryx lithographica* (HMN 1880), showing elongated mid-cervical centra (ch. 262-1). **b** right pubis of *Rahonavis ostromi* (UA 8656) in anterior view showing the presence of a shortened pubic symphysis (ch. 403-2). **c** right tibia of *Rahonavis ostromi* (UA 8656) in lateral view, showing a small lateral cnemial crest (ch. 410-1). Not to scale. *Abbreviations* cervical vertebra (*cv*), dorsal vertebra (*dv*)

376 (1) scapula relatively robust with its shaft only sharply ridged along the dorsal margin close to the distal end; 380(1) relative length of humerus subequal to or longer than femur; 386(0) ulna without a thick ridge along the anterior margin of the proximal third of the shaft; 403(2) pubic symphysis length less than 40 % total pubic length; 410(1) lateral cnemial crest of tibiotarsus small (Fig. 3.7).

Content. This node includes ((*Anchiornis* + *Xiaotingia*) + Avialae).

Comments. *Anchiornis huxleyi* was described by Xu et al. (2008) on the basis of an incomplete skeleton lacking the skull, coming from the Late Jurassic Tiaojishan Formation, of NE China. The age of these beds ranges from 161 through 150 MYR (Xu et al. 2003). *Anchiornis* was first described by Xu et al. (2008) on the basis of an incomplete skeleton lacking the skull, who considered this taxon as the basalmost member of Avialae, thus filling the gap between non-avian maniraptorans and birds. More recently, a new description based on more complete specimens of *Anchiornis huxleyi* concludes that this taxon is a basal member of Troodontidae (Hu et al. 2009; see also Xu et al. 2011b; Lee and Worthy 2011).

Recently, Xu et al. (2011a) described the non-avian theropod *Xiaotingia zhengi*, and proposed it as the sister group of *Anchiornis*, and both taxa were nested within the Archaeopterygidae, as closely related to *Archaeopteryx*. Moreover, these authors indicate that Archaeopterygidae was not directly related to the line of birds, but represented the sister group of Troodontidae and Dromaeosauridae, conforming a monophyletic Deinonychosauria. However, a detailed analysis indicates that most of the features cited by Xu and collaborators (Xu et al. 2011a) show an ambiguous distribution and others have been regarded here as convergently acquired.

Characters listed by Hu et al. (2009) as placing *Anchiornis* among Troodontidae are ambiguous, and most of them exhibit a wider distribution within paravians. Moreover, the foot of *Anchiornis* lacks troodontid attributes; on the contrary, metatarsals II and IV are subequeal in transverse diameter, and metatarsal III is not proximally embraced by metatarsals II and IV. Hu et al. (2009) refered *Anchiornis* to Troodontidae on the basis of a large maxillary fenestra separated from the antorbital fenestra by a narrow interfenestral bar (probably correlated with character 240-0 of Hu et al. 2009), a dorsoventrally flattened internarial bar (character 21-1), a distinct posteriorly widening groove on the labial surface of the dentary housing the neurovascular foramina (character 72-1), and closely packed premaxillary and dentary teeth in the symphyseal region (character 89-1) (Fig. 3.8). However, there exist some incongruence between the main text of the paper and the numerical results of its supplementary information (SI 1 of Hu et al. 2009; www.nature.com/nature). First of all, only one of the traits mentioned by the authors in the main text (the character 89-1) effectively diagnoses Troodontidae in the numerical analyses, whereas the remaining traits cited by Hu et al. (2009) did not appear as diagnostic of troodontids. However, in spite of these incongruences, we analyze here the features cited in the text, and considered by Hu et al. (2009) as uniquely shared between troodontids and *Anchiornis*. The presence of a flat internarial bar (character 21-1) is considered here as diagnostic of the clade *Anchiornis* + Avialae (e.g. *Archaeopteryx*, *Confuciusornis*;

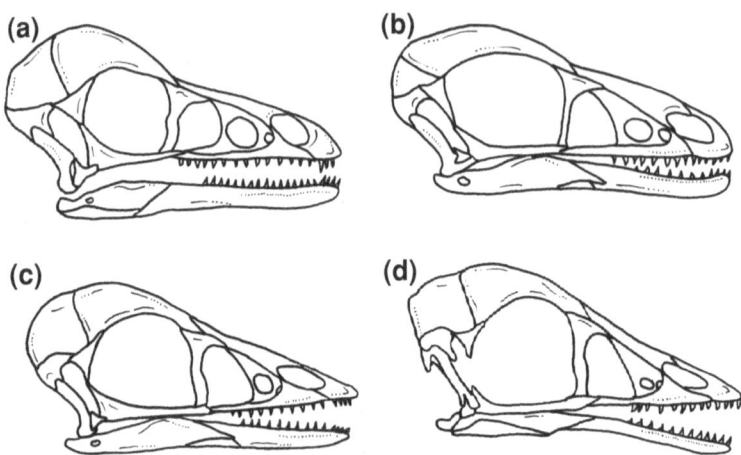

Fig. 3.8 Cranial reconstructions of *Anchiornis huxleyi* made by previous authors and comparisson with *Archaeopteryx lithographica*. **a** *Anchiornis huxleyi* based on Hu et al. (2009), **b** *Anchiornis huxleyi* based on Xu et al. 2011a, b, **c** present paper, **d** *Archaeopteryx lithographica* based on Xu et al. 2011a, b. Not to scale

Chiappe et al. 1999; Mayr et al. 2007). The presence of a lateral groove of the dentary carrying a series of neurovascular foramina (character 72-1) was analyzed by Agnolín and Novas (2011) was considered by these authors as widespread among derived maniraptorans, being more probably diagnostic of Paraves than Troodontidae. The presence of closely packed dentary and premaxillary teeth (character 89-1) was considered as diagnostic of Deinonychosauria by Xu et al. (2008) because of its presence in microraptorians. Regrettably, most unenlagiids, as well as basal birds lack well-preserved dentaries and premaxillae in which this condition may be properly known. We consider this feature as not diagnostic of Troodontidae, but of uncertain distribution among Paraves (Agnolín and Novas, 2011) . The presence of dorsal and caudal vertebrae with relatively long and slender transverse processes (character 107-1) was also considered by Hu et al. (2009) as a derived trait shared by *Anchiornis* and Troodontidae. However, on the basis of the photographs of that paper (Hu et al. 2009; SI 1, Fig. S3) it is clear that the transverse processes of *Anchiornis* are well extended anteroposteriorly, as occurs in most theropods. Proximal caudal vertebrae of *Anchiornis* clearly show rod-like transverse processes (also noted by Xu et al. 2008). Although proposed as diagnostic of Troodontidae, the presence of such elongate transverse processes on proximal caudals have been reported in a variety of basal avialans, such as *Archaeopteryx* and *Sapeornis* (Zhou and Zhang 2003a, b; Mayr et al. 2007), and are clearly present in the first caudal of the microraptorian *Sinornithosaurus* (Xu 2002) and in *Xiaotingia* (Xu et al. 2011a). Finally, the large maxillary fenestra separated from the antorbital fenestra by a narrow interfenestral bar (Hu et al. 2009) does not appear as an individual character in their numerical analysis. A single trait (character 240) is the only feature cited by Hu et al. (2009) that involves the size of this structure, which is is expressed by the authors as: "Maxillary fenestra large and round". However, a very large max-illary fenestra is not unique to troodontids, being also present in *Buitreraptor* and *Archaeopteryx* (Fig. 3.8). In conclusion, the features employed by Hu et al. (2009) are more widespread than previously thought, and most of them are also present in basal Avialae, such as *Archaeopteryx*.

Besides, there are several derived features that unite *Anchiornis* with birds, thus supporting its inclusion as stem Avialae, as originally proposed by Xu et al. (2008).

On the other hand, the present analysis departs from the proposal of Xu et al. (2011a) that *Anchiornis* and *Xiaotingia* are archaeopterygids, and the Archaeopterygidae sensu Xu et al. (2011a) is considered as a paraphyletic assem-blage, with the clade *Anchiornis + Xiaotingia* constituting the sister group of the clade formed by *Archaeopteryx* and more derived birds. This topology is sustained by a large number of characters shared by *Archaeopteryx* and more derived birds, that are lacking in *Anchiornis* and *Xiaotingia*. A similar conclusion was also reached by Lee and Worthy (2011) based on statistical support. These authors indicate that *Anchiornis* and *Xiaotingia* are the sister groups of the remaining deinonychosaurs, whereas *Archaeopteryx* represents the basalmost bird, a conclu-sion also reached here.

Among the list of synapomorphies cited in the diagnosis of the clade (*Anchi ornis + Xiaotingia*) + Avialae, there are some of them that deserve some addi-tional comments. Novas and Puerta (1997) suggested that the most relevant dif-ferences in the theropod line to birds have more to do with changes in skeletal proportions, than anatomical novelties. In this way, *Anchiornis* as well as more derived theropods shows skeletal proportions that are clearly related to flight capabilities. In this way, in *Anchiornis* the humerus is longer than the femur, being a derived condition reminiscent of Avialae (Zhang et al. 2008). Moreover, as also occurs in basal birds, such as *Archaeopteryx* and *Jeholornis*, in *Anchiornis* the total forelimb length is at least 80 % of hindlimb length, the humerus is at least as wide as the femur, and its hand is about 130 % of the femoral length (Xu et al. 2008). In the same way, Xu et al. (2008, 2011a) considered that long and robust forelimbs are considered as diagnostic of Paraves. These authors indicate that the lengthening and thickening of forelimbs indicate a dramatic shift in fore-limb function at the base of Paraves, which may be related to aerodynamic capa-bilities. However, as pointed out before, this analysis indicates that the increase in robustness and total length of the forelimbs occurred at the common ancestor of Averaptora, and some additional flying-related traits, such as an increase in the robustness in the scapula and humeral elongation occurred at the (*Anchiornis + X iaotingia*) + Avialae clade. On the other hand, basal Paraves such as Troodontidae and Dromaeosauridae still lack most of these flying-related traits (Xu et al. 1999, 2000, 2008, 2011a).

Although we are in agreement that the changes in skeletal proportions were rather important at the basal stages of flight, there are other anatomical details that deserve comment. As noted by Xu et al. (2008), in *Anchiornis* the carpal troch-lea is more derived than in dromaeosaurids in showing a remarkable convexity on its proximal end, a condition that may have allowed this taxon a considerable abduction of the hand, probably correlated with wing-folding movements typical of the avian wrist (Xu et al. 2008). Regrettably, the carpals in unenlagiids remain unknown, and thus, this peculiar morphology of the carpal trochlea may be more widespread than depicted here.

3.6 *Anchiornis* + *Xiaotingia* clade

Synapomorphies. 52(1), 71(0), 89(1), 133(1), 178(2), 206 (1), 369(2).

Content. This clade includes the genera *Anchiornis* and *Xiaotingia*.

Comments. The genus *Xiaotingia* was described by Xu et al. (2011a) on the basis of a nearly complete specimen from the Late Jurassic Tiaojishan Formation (Xu et al. 2011a). These authors considered *Xiaotingia* as nearly related to *Anchiornis* and *Archaeopteryx*, and the three genera were included within the Archaeopterygidae Huxley 1871. Moreover, Xu et al. (2011a) proposed that the Archaeopterygidae was not nearly related to the remaining birds, but to the Deinonychosauria. Xu et al. (2011a) indicated several features that may allow including archaeopterygids within Deinonychosauria. Among these traits they mentioned: (1) large promaxillary fenestra, (2) T-shaped lacrimal, (3) a lateral longitudinal groove that expands posteriorly in the dentary, (4) manual phalanx IV-2 shorter than IV-1, (5) short ischium, (6) ischium with distally located obturator process as well as posterodistal process, and (7) extensible pedal phalanx II-2. These traits are recovered here as diagnostic of more inclusive nodes (i.e. Paraves), and some of them (e.g. characters 1, 2, 4, 6) are clearly present in several basal birds (e.g. *Confuciusornis, Sapeornis, Jeholornis*; Chiappe et al. 1999; Zhou and Zhang 2002, 2003a, b; Lee and Worthy 2011; O'Connor et al. 2011). Moreover, Xu et al. (2011a) indicate that archaeopterygids, including *Xiaotingia*, resembles deinonychosaurians, rather than birds in having a subtriangular skull profile produced by a shallow snout and expanded postorbital region (see skull reconstructions on Fig. 3.8). Xu et al. (2011) also interpreted that basal birds, scansoriopterygids and oviraptorosaurians, resemble each other in having a short and tall skull with a very deep premaxilla and smaller orbits; in particular basal birds and oviraptorosaurians were found similar in having robust and deep mandibles with large mandibular fenestra (Xu et al. 2011a). Nevertheless, all these features are regarded in the present analysis as convergently acquired in Scansoriopterygidae, Oviraptorosauria, and basal birds. Although *Sapeornis* appears to have a stout skull, this condition is not sure in *Jeholornis*, because the described specimens exhibit highly distorted skulls, which show ambiguous cranial contour and proportions. Moreover, in the remaining basal birds including *Confuciusornis, Zhongornis, Archaeorhynchus, Yanornis,* and Enantiornithes the skull contour and bone proportions are rather similar to that of *Archaeopteryx, Anchiornis, Xiaotingia,* and "deinonychosaurs" (Chiappe et al. 1999; Chiappe and Walker 2002; Zhou and Zhang 2005, 2006). In this way, the present analysis does not support the proposal that robust skulls plesiomorphic for birds (see also Lee and Worthy 2011).

Xu et al. (2011a) included *Xiaotingia* within Archaeopterygidae on the basis of (1) manual phalanx II-1 more than twice as long as III-1 (character 292.1 of Xu et al. 2011a), (2) manual phalanx III-3 much longer than III-1 and III-2 combined (character 302.2 of Xu et al. 2011a), (3) furcula with L-shaped cross-section (character 369.1 of Xu et al. 2011a), and (4) ventral notch between obturator process and ischial shaft (character 307.0 of Xu et al. 2011a). Regarding the first character, in the data matrix of Xu et al. (2011a), the double length of phalanx II-1 compared with that of III-1, is a derived condition present in *Archaeopteryx*, *Wellnhoferia*, *Confuciusornis*, *Sapeornis*, and *Anchiornis*, but is absent in *Xiaotingia* (Xu et al. 2011a). Regarding characters 2 and 3, although these are present in *Xiaotingia* and *Archaeopteryx*, both features are absent in the *Xiaotingia*'s sister group *Anchiornis*, suggesting an ambiguous distribution of these characters (Xu et al. 2008, 2011a). The fourth character is of ambiguous distribution. In fact, the distal end of the ischium of *Archaeopteryx* clearly exhibits a distal notch between the obturator process and ischial shaft. However, this condition cannot be properly observed in *Xiaotingia*, in which the distal end of ischium is not completely preserved; moreover, the preserved portion of the bone suggests that a distal notch was probably absent. In the same way, the preserved distal ischium of *Anchiornis* clearly lacks a distal notch, and its morphology is very similar to that of other averaptorans, including *Buitreraptor* and *Sinornithosaurus* (Xu et al. 1999; Makovicky et al. 2005). In this way, we consider that most characters employed by Xu et al. (2011a) in order to include *Xiaotingia* and *Anchiornis* within Archaeopterygidae are ambiguous, at least. On the contrary, in the present analysis, *Anchiornis* and *Xiaotingia* are considered as conforming a monophyletic clade which is the sister group of Avialae. It must be pointed out that Xu et al. (2011a) considered *Xiaotingia* and *Anchiornis* as very near relatives, a proposal that is reinforced here.

Under their phylogenetic framework, Xu et al. (2011a) considered that a very robust and deep skull, and possibly a herbivorous diet may represent the ancestral traits of birds. However, the present analysis indicates that these features are very likely not diagnostic for Avialae, and such morphology and dietary habits were probably convergently acquired by Oviraptorosauria, Scansoriopterygidae, *Sapeornis*, and its kin (Sapeornithidae sensu Hu et al. 2010), and also probably *Jeholornis* (see also Hu et al. 2010; O'Connor et al. 2011).

3.7 Avialae Gauthier, 1986

Definition. The theropod group that includes *Archaeopteryx lithographica* and *Passer*, their most common ancestor and all of its descendants.

Synapomorphies. 75(0); 274(2); **277(3)**; 317(1); 318(1); **372 (2)**; **373 (1)**; 381(0).

Content. This node includes (*Archaeopteryx lithographica* + *Wellnhoferia grandis*) + Ornithurae.

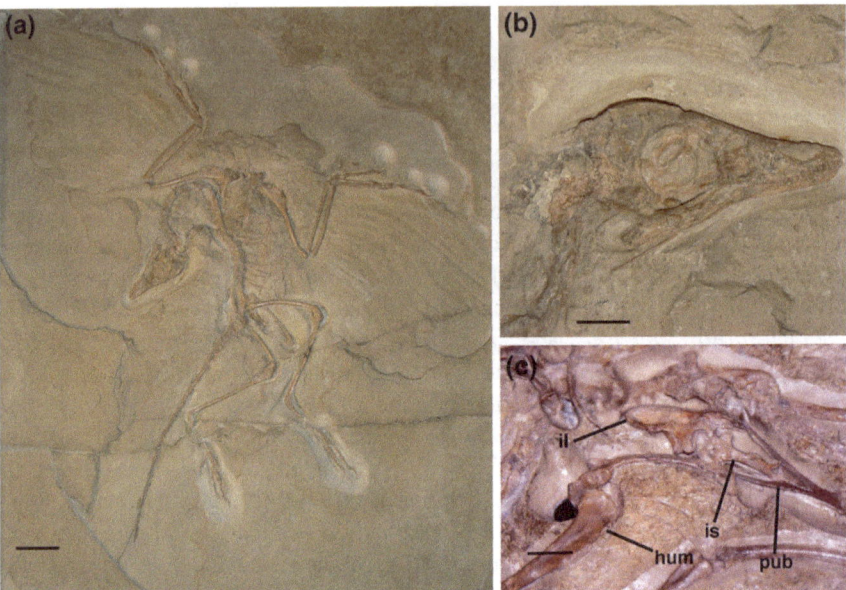

Fig. 3.9 *Archaeopteryx lithographica*. **a** London specimen mainslab (HMN 1880), **b** detailed view of the skull of the London specimen (HMN 1880), **c** Berlin specimen, detailed view of selected bones (BMNH 37001). *Abbreviations* humerus (*hum*), ilium (*il*), ischium (*is*), pubis (*pub*). Scale bar *A* 3 cm, **b–c** 5 mm

Comments. As indicated under the discussion of the clade Averaptora, we restrict here the term Avialae to the *Archaeopteryx* node, as originally proposed by Gauthier's (1986), which was followed by several authors (see Gauthier's and De Queiroz 2001; Norell et al. 2001; see Fig. 3.9). Some authors (e.g. Chiappe et al. 1999; Chiappe 2001, 2009; see Xu et al. 2011a) used the term Aves for this clade. However, following the arguments used by Gauthier's (1986; see Gauthier's and De Queiroz 2001; Clarke 2004) we restrict the term Aves for crown group birds (Neornithes sensu Chiappe 2001, 2009).

Most features diagnostic of Avialae refer to the improving of flight capabilities. In fact, the elongate humerus and radius (characters 274 and 277), as well as the extended humeral deltopectoral crest (character 381), and the strongly modified acromial portion of the scapula (characters 372 and 373) are features usually regarded as indicative of flight capabilities. In addition, archaeopterygians possessed the derived neurological adaptations required for flight (Alonso et al. 2004), but brain anatomy is still nearly unknown in most paravians, including microraptorians, unenlagiids, and *Anchiornis*.

Senter et al. (2012) proposed a tail evolutionary scenario for derived coelurosaurs, including Dromaeosauridae. They proposed that a decrease in number of caudal vertebrae occurred early in coelurosaurian phylogeny, with an increment in Eudromaeosauria. In Avialae is documented a consistent reduction in tail length from less than 24 caudal vertebrae, and a caudal/dorsosacral lenght ratio below 1.0 (Paul 2002). This pattern is increased in pygostylians, in which the tail became strongly reduced (see discussion below; Hu et al. 2010).

Arboreal habits of basal Avialae appear to have been improved with respect to more basal taxa, in having a well developed and plantar surface of hallux medially oriented and with a large claw (see Xu and Zhang 2005). This contrasts with the hypothesis of Hu et al. (2010), which proposed that an important shift towards arboreal habits can be clearly seen in Avebrevicauda (Pygostylia therein), and they suggested that at this node a significant change in locomotor system and lifestyle had occurred. Hu et al. (2010) indicated a long retroverted hallux (as implied by a medially or anteriorly oriented plantar surface of metatarsal I), phalangeal proportions, and a shortened tail (including a pygostyle). However, as indicated above, at least a partially or fully retroverted hallux was present also in *Archaeopteryx, Jeholornis*, and *Rahonavis* (Forster et al. 1998; Zhou and Zhang 2002, 2003a, b; Mayr et al. 2007; O'Connor et al. 2011). Moreover, phalangeal proportions are very similar in pygostylians, as well as more basal taxa, including *Archaeopteryx, Xiaotingia, Microraptor*, and *Jeholornis*, at least (Zhou and Zhang 2002; Mayr et al. 2007; Xu et al. 2011a). Finally, basal Avialae, including *Archaeopteryx* and *Jeholornis* began to shorten its tail, having less than 24 caudal vertebrae, a number smaller than in non-avialan theropods (Paul 2002; see below). In this way, all purported diagnostic features of arboreal lifestyle were developed at the base of Avialae, rather than Pygostylia, contrary to the hypothesis of Hu et al. (2010).

3.8 Ornithurae Haeckel, 1816

Definition. Following the phylogenetic definition of Sereno (1997), we define Ornithurae as all avialans closer to Aves than to *Archaeopteryx*.

Synapomorphies. 110 (1), 335(1), 371(1), 382(1), **401(1)**.

Content. This node includes *Rahonavis* + Avebrevicauda.

Comments. In this chapter we use the term Ornithurae as originally proposed by Gauthier's (1986), and later modified by Sereno (1997, 1999), rather than the definition by Chiappe (2001; see also Padian et al. 1999) because Gauthier's definition has priority.

In the present paper *Rahonavis* was not included within Unenlagiidae, a result that is in agreement with the original proposal by Forster et al. (1998) and the phylogenies of Zhou and Zhang (2002), Hwang et al. 2002, Novas and Pol (2005), Turner et al. (2007a), Xu et al. (2008), and Yuan (2008) (Fig. 3.10). However, the incomplete nature of the specimen, together with several similitudes with Unenlagiidae noted by previous authors (Novas 2004; Makovicky et al. 2005; Senter 2007; Novas et al. 2009; Turner et al. 2011), indicate that probably more detailed studies may reubicate *Rahonavis* within Unenlagiidae again. Several authors reported features that may diagnose the clade

Fig. 3.10 *Rahonavis*
ostromi. **a** left ilium in medial
view, **b** right pubis in lateral
view, **c** right incomplete
ischium in lateral view, **d** left
incomplete ischium in lateral
view. *Abbreviations* ischiadic
process (*ip*), obturator
process (*op*), posterodorsal
process (*pdp*). Scale bar 1 cm

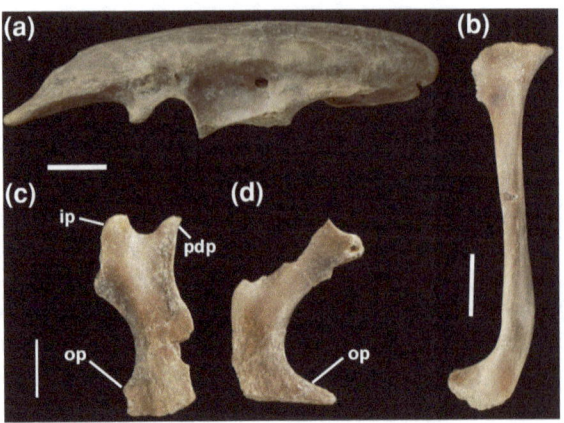

Unenlagiidae + *Rahonavis*, including: prominent supracetabular crest (character 157-1 in Makovicky et al. 2005), vertically oriented pubis (character 177-1
in Makovicky et al. 2005), dorsocaudal edge of ilium concave (character 227-1
in Makovicky et al. 2005), dorsal vertebrae with transverse process shortened
(Novas et al. 2009), dorsal vertebrae with neural spines transversely expanded
into a spine table (Novas et al. 2009), and metatarsal III proximally pinched
(Novas et al. 2009) (Fig. 3.10). However, as pointed out above, the feature 157-1
in Makovicky et al. (2005) is plesiomorphic for theropods and is highly variable
among basal birds and non-avian theropods. Nevertheless, in *Rahonavis*, as well
as *Archaeopteryx* and *Jeholornis* the supracetabular crest is poorly developed as a
thin ridge (Zhou and Zhang 2002; Novas 2004). A vertically oriented pubis (character 175-1 herein) is present in *Rahonavis*, *Unenlagia*, and *Buitreraptor* (Novas
and Puerta 1997; Forster et al. 1998; Calvo et al. 2004; Makovicky et al. 2005),
but is also present in basalmost avialans, as for example *Archaeopteryx* and
Jeholornis (Paul 2002; Zhou and Zhang 2002). The presence of a dorsocaudally
concave margin of ilium (character 227-1 in Makovicky et al. 2005) was proposed by Makovicky et al. (2005) as a synapomorphy of Unenlagiidae, an interpretation also followed by Agnolín and Novas (2011; character 332-1). However,
the presence of a concave dorsocaudal iliac margin is also seen in the averaptoran
Tianyuraptor (Zheng et al. 2009), as well as in some specimens of *Archaeopteryx*
(Novas 2004), *Sapeornis* (Zhou and Zhang 2003a, b), and *Confuciusornis*
(Chiappe et al. 1999) (Fig. 3.10). The presence of short transverse processes on
dorsal vertebrae (character 107-1 herein) is a feature plesiomorphic for theropods
being widely spread among tetanurans (Hu et al. 2009). Regarding dorsal vertebrae, the presence of a spine table at the top of the neural spine was proposed as a
synapomorphy of Unenlagiidae by Novas et al. (2009). However, a spine table is
absent in *Buitreraptor* and *Rahonavis*, and was invalidated as a unenlagiid synapomorphy by Gianechini et al. (2011a). Proximally pinched metatarsal III (character 200-1 herein) was considered as diagnostic of Unenlagiidae by Novas et
al. (2009); however, its presence in microraptorians, *Anchiornis*, and basal and

derived avialans allow to recognize this trait here as diagnostic of a more inclusive clade, that is Averaptora.

Most features diagnostic of the Ornithurae are rather weak and present an ambiguous distribution among Avialae. Moreover, there are some traits that may have been more widely distributed among theropods. For example, the character 110-1 concerns the number of sacral vertebrae. In fact, in Ornithurae there are six or more sacrals. However, in more basal taxa, including *Unenlagia* and *Velociraptor*, there are also six sacrals, resembling Ornithurae birds in this aspect (Novas and Puerta 1997; Norell and Makovicky 1999). However, as indicated by Norell and Makovicky (1999) juvenile specimens of *Velociraptor* show five sacrals, indicating that the number of sacrals probably varies with ontogeny. In addition, the number of sacrals is uncertain in the laterally compressed *Archaeopteryx* specimens and in the poorly preserved sacrum of *Rahonavis* (Forster et al. 1998). Although in the latter more than six sacrals were surely present.

Another putative character diagnostic of Ornithurae is the ulna, which is much more robust than the tibiotarsus robustness (character 382-1). However, this condition is also seen in non-Ornithurae basal taxa, including Microraptoria (e.g. *Microraptor, Sinornithosaurus*; Senter et al. 2004a), *Anchiornis* (Xu et al. 2008), and probably *Xiaotingia* (Xu et al. 2011a). In *Archaeopteryx* the ulnar robustness is still a matter of debate; while some authors (e.g. Paul 2002) indicate that the ulna is much more robust than the tibiotarsus, others (Hu et al. 2009) point out that the ulna in this taxon is thinner than the tibiotarsus, an interpretation that is followed here.

3.9 Avebrevicauda Paul, 2002

Definition. All avians closer to Aves than to *Jeholornis* or *Rahonavis*.

Synapomorphies. 110(2), **121(2)**, 134(0), 148(2), 195(1), 197(1), 201(0), 406(1).

Content. This node includes *Sapeornis* + Pygostylia.

Comments. The Avebrevicauda was originally diagnosed by Paul (2002) as an apomorphy-based clade, including birds in which the free caudals were reduced to ten or that descended from such avians (Fig. 3.11). We here redefine the concept of Paul (2002) for the sake of clarity, and we reconsider Avebrevicauda as a stem-based clade. Hu et al. (2010) recognized this clade under the name Pygostylia. However, in the definition of Pygostylia, Chatterjee (1997) considered this clade as the group that includes "the common ancestor of Confuciusornithidae and Neornithes plus all of its descendants" (Chiappe 2001). In this way, we opt to use another name for the more inclusive clade that includes Pygostylia and *Sapeornis*.

Avebrevicaudans lack the hyperextensible pedal digit II, present in most paravians, including basal birds, and once considered as diagnostic of Deinonychosauria (Gauthier's 1986). By contrast, the reduction of ungual and preungual phalanges of pedal digit II is considered here as diagnostic of the clade (character 201-0).

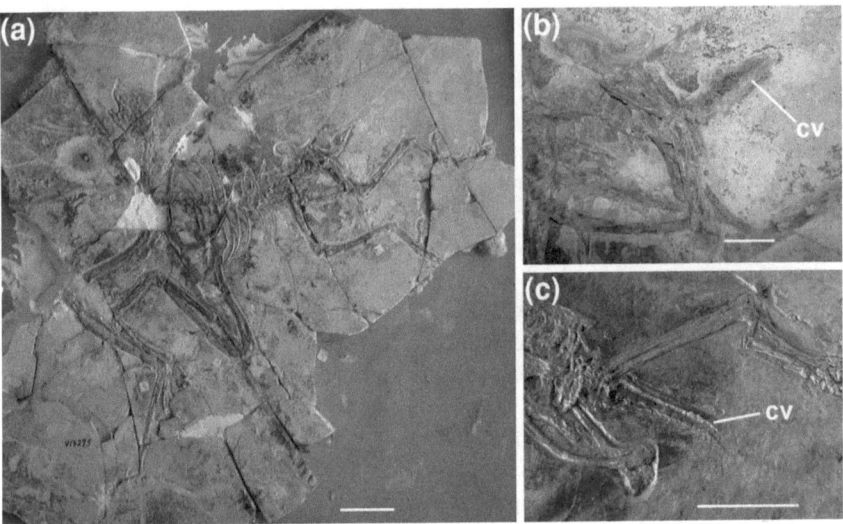

Fig. 3.11 Selected avebrevicaudans. **a, b,** *Sapeornis chaoyangensis* (IVPP V 13275), **a** mainslab of the most complete *Sapeornis* skeleton, **b** detail of reduced tail with pygostyle-like terminal vertebrae. **c** *Confuciusornis sanctus* (IVPP V 11374). *Abbreviations* cv, pygostyle-like caudal vertebrae. Scale bar **a** 5 cm, **b, c** 2 cm

Regarding caudofemoral musculature with its femoral osteological correlates Hutchinson (2001) indicates that the fourth trochanter of femur and its medial pit are correlates of the insertion of the *M. caudofemoralis longus*, and the trochanteric shelf seems to be a direct correlate of the *M. iliofemoralis* insertion, especially the *M. iliofemoralis externus* of Aves (Avialae herein). The laterally extended trochanteric shelf seen in dromaeosaurids may indicate a slight expansion of the *M. caudofemoralis longus* which was then reduced in basal Ornithothoraces to the tiny *M. caudofemoralis longus* seen in Neornithes (Aves herein; see Gauthier 1986; Gauthier and De Queiroz 2001). Thus, as also indicated by Gatesy (1990), the role of the *M. caudofemoralis longus* as a femoral abductor was reduced together with the trochanteric crest of femur. However, it is worth mention that the absence of a fourth trochanter on femur is also reported in basal averaptorans (e.g. *Microraptor, Sinornithosaurus, Unenlagia, Buitreraptor, Austroraptor*; Xu 2002; Makovicky et al. 2005; Novas and Pol 2005; Novas et al. 2009), *Xiaotingia* (Xu et al. 2011a), and *Anchiornis* (Xu et al. 2008). Moreover, a reduced trochanteric shelf is not restricted to avebrevicaudan birds, but also in more basal taxa (e.g. *Buitreraptor, Archaeopteryx*; MPCA 245; HMN 1880), in which this shelf is represented only by a very small, low, and rounded bump (Agnolín and Novas 2011). Both features indicate that basal averaptorans possessed a reduced *M. iliofemoralis externus* and *M. caudofemoralis longus*. Moreover, a pit for the small but prominent *M. iliotrochantericus caudalis* inserted on the craniolateral rim of the trochanteric bump. This condition was considered

Fig. 3.12 Proximal end of
left femur of *Buitreraptor
gonzalezorum* (MPCA 245)
showing the attachment of
the *M. iliotrochantericus
caudalis* (*Mic*) and the
trochanteric shelf (*TS*)

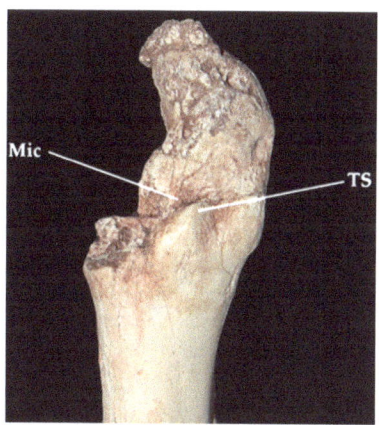

as diagnostic of Ornithothoraces by Hutchinson (2001), but is clearly present at least in *Buitreraptor*, and was also probably present in *Archaeopteryx* (Fig. 3.12).

As a concluding remark, we may point out that the reduction in caudal musculature as typical of modern birds with hyper abbreviated tails was also present in basal long-tailed taxa, suggesting that morphological changes in such muscles were probably well in progress before the reduction of the tail.

3.10 Oviraptorosauria + Scansoriopterygidae clade

Synapomorphies. 20(0) maxillary process of premaxilla contacts nasal to form posterior border of nares; 33(1) jugal quadratojugal process rodlike; 39(1) enlarged foramen opening laterally at the angle of the lacrimal; **68(1)** mandible with coronoid prominence; 69(1) posterodorsal process above anterior end of mandibular fenestra; 74(1) internal mandibular fenestra large and rounded; **79(1)** retroarticular process elongate and slender; 255(1) dentary with posteroventral process extending to posterior end of external mandibular fenestra; 269(1) acromion process of scapula reduced and does not contact coracoid; 288(0) length of manual phalanx II-2 < 1.2 × length of phalanx II-1; 427(1) Main axis of external naris subvertical; 428(1) parietal nuchal transverse crest absent (Fig. 3.13).

Content. This node includes Scansoripterygidae (sensu Zhang et al. 2008) + Oviraptorosauria.

Comments. This family of bizarre coelurosaurians includes *Epidexipteryx* and *Epidendrosaurus*, from the Middle to Late Jurassic Daohugou sediments of Inner Mongolia, China (Zhang et al. 2002, 2008). Scansoriopterygids are small-sized taxa (approximately 26 cm long from the tip of bony tail to the premaxilla) characterizaed by the elongation of manual digit III, which considerably exceeds the length of the remaining digits. Zhang et al. (2002, 2008) proposed scansoriopterygids as the basalmost representatives of Avialae, a criterion followed by other authors

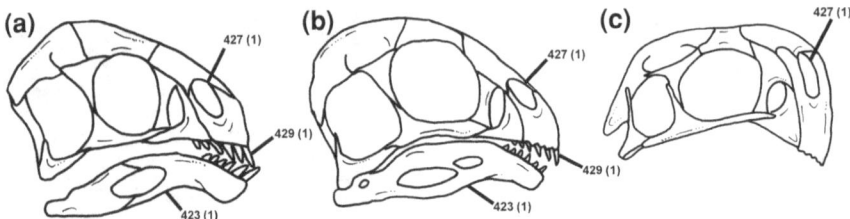

Fig. 3.13 Cranial reconstruction of Oviraptorosauria and Scansoriopterygidae. **a** the scansorio-pterygid *Epidexipteryx hui* modified from Zhang et al. (2008) and Xu et al. (2011a, b), **b** the ovi-raptorosaurian *Similicaudipteryx yixianensis* modified from Xu et al. (2011a, b), **c** *Conchoraptor gracilis* based on Osmolska et al. (2004). Not to scale

(e.g., Xu and Zhang 2005; Hu et al. 2009; Chioiniere et al. 2010; Xu et al. 2011a). Several features have been identified by Zhang et al. (2002, 2008) in support of their phylogenetic interpretation, but most of them are conflictiing (see below).

Zhang et al. (2008) made a numerical analysis in order to sustain the sister-group relationship between Scansoriopterygidae and Avialae. However, features recog-nized by these authors in support of this phylogenetic hypothesis deserve the fol-lowing observations: (1) dentary and maxillary teeth devoid of serrations (character 84-0) is not uniquely present in avialans, but is also documented in *Incisivosaurus*, *Caudipteryx*, and Alvarezsauridae (Chioiniere et al. 2010); (2) dorsal vertebrae with parapophyses flushing with the neural arch (character 103-0) constitute a plesiomor-phic condition among theropods (Norell and Makovicky 2004; Agnolín and Novas 2011); (3) reduction in the number of caudals (that is, presence of 21-30 tail verte-brae; character 121-1) is not only present among basal Avialae, but is also docu-mented in oviraptorans, in which less than 32 caudal vertebrae are present (e.g., 22 caudals in *Caudipteryx*, and 24 in *Nomingia*; Osmolska et al. 2004); (4) presence of a metatarsal cap with intercondylar eminence fused to metatarsals (character 196-1), is a condition also seen in Alvarezsauria (Chioiniere et al. 2010) and in the oviraptoran *Avimimus* (Osmolska et al. 2004); (5) Zhang et al. (2008) indicated the presence of a shallow meckelian groove of dentary as a synapomorphy uniting both groups, but the medial surface of the dentary remains unknown in Scansoriopterygidae (Zhang et al. 2002; 2008); (6) humerus longer than femur is listed by Zhang et al. (2008) as a derived condition shared by scansoriopterygids and avialans, but this character is optimized in the present paper as diagnostic of Averaptora (see above; character 273-2); (7) Zhang et al. (2008) suggested the presence of a reversed hallux (charac-ter 317-1) as shared by scansoriopterygids and Avialae, but in *Epidendrosaurus* (the only known scansoriopterygid in which the hallux is properly preserved) the hallux is laterally oriented and its plantar surface is posteriorly directed, as plesiomorphi-cally occurs in most theropods (Zhang et al. 2002); (8) presence of an unreduced ungual phalanx in the hallux (character 318-1) is a valid feature shared by scanso-riopterygids and avialans, but this condition is optimized in the present paper as diagnostic of Paraves (see above; Agnolín and Novas 2011); (9) pedal phalanx II-2 longer than phalanx II-1 listed for scansoriopterygids and avialans, is also present in *Caudipteryx* and *Avimimus*, among other oviraptorans (Kurzanov 1981; Osmolska et al. 2004); (10) character 158 (state 0) is included in the list of synapomorphies of Scansoriopterygidae plus Avialae, but this feature is not defined in the character list in Zhang et al. (2008; SI), thus it cannot be properly analyzed here; (11) preacetabular process of ilium elongate and with a strongly convex cranial margin has also been cited as an avialan feature of scansoriopterygids, but this feature is not included in the numerical analysis made by Zhang et al. (2008).

Zhang et al. (2008) reported several similitudes between scansoriopterygids and oviraptorans that are worth of mention: skull short and high, external nares posited high on the snout, mandible downturned and dorsally convex, mandibular fenestra large, and anterior teeth cylindrical and procumbent (Fig. 3.13). Moreover, the already cited avian-like condition for the preacetabular wing of ilium (being elongate and with a strongly convex cranial margin) is also seen in basal oviraptorans, such as *Nomingia* and *Caudipteryx* (Osmolska et al. 2004); in this regard, the ilium of *Epidexipteryx* (Zhang et al. 2008) resembles that of the oviraptorosaurian *Avimimus* in being elongate and dorsoventrally depressed and in having an elongate, rounded and low preacetabular wing, a combination of traits not seen in other theropod dinosaurs (Kurzanov 1981). In this way, Xu et al. (2010a) suggested a sister-group relationship between Oviraptorosauria + Scansoriopterygidae, and this clade resulted as the sister group of Avialae; however, Xu et al. (2010a) did not provide a numerical analysis or formal characters in order to sustain their hypothesis.

In agreement with the suggestion of Xu et al. (2011a), present cladistic analysis support, contrary to the avialan relationships proposed by previous authors (Zhang et al. 2002, 2008; Xu and Zhang 2005; Zhang et al. 2008; Hu et al. 2009; Chioiniere et al. 2010), that scansoriopterygids are basal maniraptorans which exhibits sister group relationships with Oviraptorosauria.

References

Agnolín FL, Novas FE (2011) Unenlagiid theropods: are they members of dromaeosauridae (theropoda, maniraptora). An Acad Bras Ciênc 83:117–162

Alonso PD, Milner AC, Ketcham RA, Cokson MJ, Rowe TB (2004) The avian nature of the brain and inner ear of *Archaeopteryx*. Nature 430:666–669

Barsbold R, Osmólska H (1999) The skull of *Velociraptor* (Theropoda) from the Late Cretaceous of Mongolia. Acta Palaeont Pol 44:189–219

Bonaparte JF (1999) Tetrapod faunas from South America and India: a palaeobiogeographic interpretation. Proc India Acad Sci 65:427–437

Burnham D, Currie PA, Bakker R, Zhou Z, Ostrom J (2000) Remarkable new birdlike dinosaur (Theropoda: Maniraptora) from the Upper Cretaceous of Montana. Univ Kansas Paleont Contrib 13:1–14

Calvo JO, Porfiri JD, Kellner AWA (2004) On a new maniraptoran dinosaur (Theropoda) from the Upper Cretaceous of Neuquén, Patagonia, Argentina. Arquiv Mus Nac 62:549–566

Campbell KE (2008) The manus of archaeopterygians: implications for avian ancestry. Oryctos 7:13–26

Carpenter K (2002) Forelimb biomechanics of nonavian theropod dinosaurs in predation. Senck Leth 82:59–76

Chiappe LM (2001) Phylogenetic relationships among basal birds. In: Gauthier's J, Gall LF (eds) New perspectives on the origin and evolution of birds. Yale University, New Haven, pp 125–139 (Special Publications of the Peabody Museum of Natural History)

Chiappe LM (2009) Downsized dinosaurs: the evolutionary transition to modern birds. Evo Edu Outreach 2:248–256

Chiappe LM, Walker CA (2002) Skeletal morphology and systematics of Cretaceous euenantiornithes (Ornithothoraces: Enantiornithes). In: Chiappe LM, Witmer LM (eds) Mesozoic birds: above the heads of dinosaurs. Berkeley University Press, Berkeley, pp 168–218

Chiappe LM, Ji S, Ji Q, Norell MA (1999) Anatomy and systematics of Confuciusornithidae from the Late Mesozoic of Northeastern China. Bull Amer Mus Nat Hist 242:1–89

Choiniere JN, Xu X, Clark JM, Forster CA, Guo Y, Han F (2010) A basal alvarezsauroid theropod from the early Late Jurassic of Xinjiang, China. Science 327:571–574

Clarke J (2004) Morphology, phylogenetic taxonomy, and systematics of *Ichthyornis* and *Apatornis* (Avialae: Ornithurae). Bull Am Mus Nat Hist 286:1–179

Colbert EH, Russell DA (1969) The small Cretaceous dinosaur Dromaeosaurus. Amer Mus Novitates 2380:1–49

Currie PJ (1985) Cranial anatomy of *Stenonychosaurus inequalis* Saurischia, Theropoda and its bearing on the origin of birds. Canadian J Earth Sci 22:1643–1658

Currie PJ (1987) Bird-like characteristics of the jaws and teeth of troodontid theropod (Dinosauria, Saurischia). J Vert Paleont 7:72–81

Currie PJ (2000) Theropods from the Cretaceous of Mongolia. The Age of Dinosaurs in Russia and Mongolia. Cambridge University Press, Cambridge, pp 434–455

Forster C, Sampson S, Chiappe LM, Krause D (1998) The theropod ancestry of birds: new evidence from the Late Cretaceous of Madagascar. Science 279:1915–1919

Gatesy SM (1990) Caudifemoralis musculature and the evolution of theropod locomotion. Paleobiology 16:170–186

Gauthier's JA (1986) Saurischian monophyly and the origin of birds. Mem Calif Acad Sci 8:1–46

Gauthier's J, de Queiroz K (2001) Feathered dinosaurs, flying dinosaurs, crown dinosaurs, and the name "Aves". In: Gauthier J, Gall LF (eds) New perspectives on the origin and early evolution of birds. In: Proceedings of the international symposium in honor of John H. Ostrom. Peabody Museum of Natural History, Yale University, New Haven, Connecticut, pp 7–41

Gianechini FA, Apesteguía S (2011) Unenlagiinae revisited: dromaeosaurid theropods from South America. An Acad Bras Ciênc 83:163–197

Gianechini FA, Apesteguía S, Makovicky PJ (2011a) New information on the cranial and vertebral anatomy of *Buitreraptor gonzalezorum* (theropoda: unenlagiinae) and preliminary comparissons with other South American unenlagiines. Ameghiniana, vol 52 (in press)

Hu D, Hou L, Zhang L, Xu X (2009) A pre-*Archaeopteryx* troodontid theropod from China with long feathers on the metatarsus. Nature 461:640–643

Hu D, Li L, Hou L, Xu X (2010) A new sapeornithid bird from China and its implications for early avian evolution. Act Geol Sin 84:472–482

Hutchinson JR (2001) The evolution of pelvic osteology and soft tissues on the line to extant birds (Neornithes). Zool J Linn Soc 131:123–168

Hwang SH, Norell MA, Qiang J, Keqin G (2002) New specimens of *Microraptor zhaoianus* (Theropoda: Dromaeosauridae) from Northeastern China. Amer Mus Novit 3381:1–44

Kurzanov SM (1981) An unusual theropod from the Upper Cretaceous of Mongolia. Joint Soviet-Mongolian Paleontological Expedition 15:39–49

Lee MS, Worthy TH (2011) Likelihood reinstates *Archaeopteryx* as a primitive bird. Biology Letters doi:10.1098/rsbl.2011.0884

Longrich NR, Currie PJ (2008) A microraptorine (Dinosauria-Dromaeosauridae) from the Late Cretaceous of North America. PNAS 106:5002–5007

Makovicky PJ, Apesteguía S, Agnolín FL (2005) The earliest dromaeosaurid theropod from South America. Nature 437:1007–1011

Mayr G, Pohl B, Hartman S, Peters DS (2007) The tenth skeletan specimen of *Archaeopteryx*. Zool Jour Lin Soc 149:97–116

Norell MA, Makovicky PJ (1999) Important features of the dromaeosaur skeleton II: information from newly collected specimens of *Velociraptor mongoliensis*. Amer Mus Novit 3282:1–45

Norell MA, Makovicky PJ (2004) Dromaeosauridae. In: Weishampel DB, Dodson P, Osmolska H (eds) The dinosauria, 2nd edn. University of California Press, Berkeley, pp 196–209

Norell MA, Clark JM, Makovicky PJ (2001) Phylogenetic relationships among coelurosaurian theropods. In: Gauthier J, Gall LF (eds) New perspectives on the origin and early evolution of birds. Peabody Museum of Natural History, Yale University, New Haven, pp 49–68

Norell MA, Makovicky PJ, Clark JM (2000) A new troodontid from Ukhaa Tolgod, Late Cretaceous, Mongolia. J Vert Paleontol 20:7–11

Novas FE (1999) Dinosaur. McGraw-Hill Yearbook of Science & Technology, pp 133–135

Novas FE (2004) Avian traits in the ilium of *Unenlagia comahuensis* (Maniraptora, Avialae). In: Currie PJ, Koppelhus EB, Shugar MA, Wright JL (eds) Feathered dragons: studies on the transition from dinosaurs to birds. Indiana University Press, Bloomington, pp 150–166

Novas FE (2009) The Age of Dinosaurs in South America. Indiana University Press, Indiana, p 452

Novas FE, Pol D (2005) New evidence on deinonychosaurian dinosaurs from the Late Cretaceous of Patagonia. Nature 433:858–861

Novas FE, Puerta P (1997) New evidence concerning avian origins from the Late Cretaceous of NW Patagonia. Nature 387:390–392

Novas FE, Pol D, Canale JI, Porfiri JD, Calvo JO (2009) A bizarre Cretaceous theropod dinosaur from Patagonia and the evolution of Gondwanan dromaeosaurids. Proc Royal Soc Lon B 126:1101–1107

O'Connor JK, Sun C, Xu X, Wang X, Zhou Z (2011) A new species of *Jeholornis* with complete caudal integument. Hist Biol 2011, First article

Ostrom JH (1969) Osteology of *Deinonychus antirrhopus*, an unusual theropod from the Lower Cretaceous of Montana. Bull Peabody Mus Nat Hist 30:1–165

Ostrom JH (1976) *Archaeopteryx* and the origin of birds. Biol J Linn Soc 8:91–182

Padian K, Ricqlès A (2009) L'origine et l'évolution des oiseaux: 35 années du progrès/The origin and evolution of birds: 35 years of progress. Comptes Rendus PalEvol 8:257–280

Padian K, Hutchinson RM, Holtz TR (1999) Phylogenetic definitions and nomenclature of the major taxonomic categories of the carnivorous dinosauria (theropoda). J Vert Paleont 19:69–80

Paul GS (2002) Dinosaurs of the air. The John Hopkins University Press

Perle A, Norell MA, Chiappe LM, Clark JM (1993) Flightless bird from the Cretaceous of Mongolia. Nature 362:623–626

Porfiri JD, Calvo JO, Dos Santos D (2011) A new small deinonychosaur (Dinosauria: Theropoda) from the Late Cretaceous of Patagonia, Argentina. An Acad Bras Ciênc 83:109–116

Rauhut OWM (2003) The interrelationships and evolution of basal theropod dinosaurs. Spec Pap Palaeont 69:1–213

Senter P (2007) A new look at the phylogeny of Coelurosauria (Dinosauria: Theropoda). J Syst Palaeont 5:429–463

Senter P, Barsbold R, Britt B, Burnham D (2004) Systematics and evolution of Dromaeosauridae (Dinosauria: Theropoda). Bull Gunma Mus Nat Hist 8:1–20

Sereno PC (1997) The origin and evolution of dinosaurs. Ann Rev Earth Planet Sci 25:435–489

Sereno PC (1998) A rationale for phylogenetic definitions, with application to the higher level taxonomy of Dinosauria. N Jahr Geol Paläont 210:41–83

Sereno PC (1999) The evolution of dinosaurs. Science 284:2137–2147

Turner AH, Pol D, Clarke JA, Erickson GM, Norell MA (2007a) A basal dromaeosaurid and size evolution preceding avian flight. Science 317:1378–1381

Turner AH, Hwang SH, Norell MA (2007b) A small derived theropod from Oösh, Early Cretaceous, Baykhangor Mongolia. Amer Mus Novit 3557:1–27

Turner AH, Makovicky PJ, Norel MA (2007c) Feather quill knobs in the dinosaur *Velociraptor.* Science 21317:1721

Turner AH, Pol D, Mark AN (2011) Anatomy of *Mahakala omnogovae* (Theropoda:Dromaeosauridae), Tögrögiin Shiree, Mongolia. Amer Mus Novitates 3722:1–66

Xu X (2002) Deinonychosaurian fossils from the Jehol Group of western Liaoning and the coe-lurosaurian evolution. Dissertation for the Doctoral Degree, Chinese Academy of Sciences, Beijing

Xu X, Norell MK (2004) A new troodontid dinosaur from China with avian-like sleeping pos-ture. Nature 431:838–841

Xu X, Wang X-L (2004) A A new dromaeosaur (Dinosauria: Theropoda) from the Early Cretaceous Yixian Formation of western Liaoning. Vert Palas 42:11–119

Xu X, Zhang F (2005) A new maniraptoran dinosaur from China with long feathers on the meta-tarsus. Naturwissenschaften 92: 173–177

Xu X, Wang X-L, Wu X-C (1999) A dromaeosaurid dinosaur with a filamentous integument from the Yixian Formation of China. Nature 401:262–266

Xu X, Zhou Z, Wang X (2000) The smallest known non-avian theropod dinosaur. Nature 408:705–708

Xu X, Zhou Z, Wang X, Huang X, Zhang F, Du X (2003) Four winged dinosaurs from China. Nature 421:335–340

Xu X, Zhao Q, Norell MA, Sullivan C, Hone D, Erickson PG, Wang X, Han F, Guo Y (2008) A new feathered dinosaur fossil that fills a morphological gap in avian origin. Chinese Sci Bull 54:430–435

Xu X, Ma QY, HU DY (2010) Pre-*Archaeopteryx* coelurosaurian dinosaurs and their implica-tions for understanding avian origins. Chinese Sci Bull 55:3971–3977

Xu X, You H, Du K, Han F (2011a) An *Archaeopteryx*-like theropod from China and the origin of Avialae. Nature 475:465–470

Xu X, Tan Q, Sullivan C, Han F, Xiao D (2011b) A short-armed troodontid dinosaur from the Upper Cretaceous of Inner Mongolia and its implications for troodontid evolution. PlosOne 6:e22916

Yuan C (2008) A new genus and species of Sapeornithidae from Lower Cretaceous in western Liaoning, China. Acta Geol Sinica 82:48–55

Zanno LE, Varricchio DJ, O´Connor PM, Titus AL, Knell MJ (2011) A new troodontid theropod, *Talos sampsoni* gen. et sp. nov., from the Upper Cretaceous Western Interior Basin of North America. PlosOne 6:e24487

Zhang F, Zhou Z (2000) A primitive enantiornithine bird and the origin of feathers. Science 290:1955–1959

Zhang F, Zhou Z, Xu X, Wang X (2002) A juvenile coelurosaurian theropod from China indi-cates arboreal habits. Naturwissenschaften 89:394–398

Zhang F, Zhou Z, Xu X, Wang X, Sullivan C (2008) A bizarre Jurassic maniraptoran from China with elongate ribbon-like feathers. Nature 455:1105–1108

Zheng X, Xu X, You H, Zhao Q, Dong Z (2009) A short-armed dromaeosaurid from the Jehol Group of China with implications for early dromaeosaurid evolution. Proc Royal Soc London B 277:211–217

Zhou ZH, Zhang FC (2002) A long-tailed, seed-eating bird from the Early Cretaceous of China. Nature 418:405–409

Zhou ZH, Zhang FC (2003a) Anatomy and systematics of the primitive bird *Sapeornis chaoyan-gensis* from the early Cretaceous of Liaoning, China. Canadian J Earth Sci 40:731–747

Zhou ZH, Zhang FC (2003b) *Jeholornis* compared to *Archaeopteryx*, with a new understanding of the earliest avian evolution. Naturwissenschaften 90:220–225

Zhou Z, Zhang F (2005) Discovery of an ornithurine bird and its implication for Early Cretaceous avian radiation. PNAS 102:18998–19002

Zhou Z, Zhang F (2006) A beaked basal ornithurine bird (Aves, Ornithurae) from the Lower Cretaceous of China. Zool Script 35:363–373

Chapter 4
Uncertain Averaptoran Theropods

4.1 Introduction

Several theropods have been briefly described and assigned to the clade Dromaeosauridae. However, some of these taxa may be excluded from such theropod group, based on several osteological features. Regrettably, most of these taxa were only briefly described, or are rather fragmentary, and consequently a complete cladistic analysis is beyond the present article. However, some comments about salient features of each of these taxa are here performed in order to determine their possible phylogenetic positions:

4.2 *Luanchuanraptor henanensis*

This taxon was described by Lu et al. (2007) on the basis of a poorly preserved skeleton from the Late Cretaceous of China. These authors included *Luanchuanraptor* within Dromaeosauridae on the basis of teeth without constriction at base, stalked parapophyses on dorsal vertebrae, and elongate caudal prezygapophyses. However, as previously discussed by Agnolín and Novas (2011), teeth without constrictions at base is a widespread plesiomorphic condition among theropods and is not diagnostic of Dromaeosauridae. Moreover, the presence of stalked parapophyses was also considered by Agnolín and Novas (2011) as diagnostic of Paraves, rather than Dromaeosauridae. Regarding elongate prezygapophyses of caudal vertebrae, Lu et al. (2007) remarked that in *Luanchuanraptor* each prezygapophyses is shorter than in remaining Dromaeosauridae, thus, this theropod shows the typical condition seen in most paravians, in which prezygapophyses span less than half of the preceding vertebra. In the same way, *Luanchuanraptor* differs from dromaeosaurids and resembles averaptorans in the enlarged deltopectoral crest on humerus, cervical epipophyses shorter than postzygapophyses (see Agnolín and Novas 2011), and ilium without supracetabular crest and reduced antitrochanter (see above; Burnham 2008). This combination of traits allow us to assign *Luanchuanraptor* to Averaptora. Moreover, this genus shows a large fenestra on the coracoid, a synapomorphic condition of Microraptoria (Zheng et al. 2009), suggesting the assignment of *Luanchuanraptor* to that clade.

F. L. Agnolín and F. E. Novas, *Avian Ancestors,* SpringerBriefs in Earth System Sciences, 37
DOI: 10.1007/978-94-007-5637-3_4, © The Author(s) 2013

4.3 *Hulsanpes perlei*

It was described by Osmólska (1982) on the basis of a single and incomplete foot from the Latest Cretaceous of Mongolia. This taxon was referred with doubts to Dromaeosauridae by Osmólska (1982), a criteria followed by Norell and Makovicky (2004). On the other hand, Currie (2000) indicated that on the basis of the absence of dromaeosaurid apomorphies in the type and only known specimen, *Hulsanpes* may be excluded from Dromaeosauridae, and this author suggested that it may belong to another raptor-like clade. In fact, *Hulsanpes* differs from dromaeosaurids in having poorly excavated distal end of metatarsals II, III, and IV lacking of a ginglymoid articular end (Norell and Makovicky 2004). Moreover, phalanx 2-II lacks the extensive posteroventral heel typical of dromaeosaurids, being craniocaudally shorter, as occurs in basal birds (e.g. *Jeholornis*, *Archaeopteryx*; Paul 2002; Agnolín and Novas 2011). Moreover metatarsal III is proximally pinched, a diagnostic trait of Averaptora (see above), and its metatarsals are extremely gracile, as occurs in most Avialae (Xu and Zhang 2005). In this way, the morphology of the foot of *Hulsanpes* suggests its exclusion from Dromaeosauridae, being here considered as an uncertain Averaptora.

4.4 *Shanag agile*

It comes from the Lower Cretaceous of Mongolia based on a single specimen that includes an incomplete maxilla and dentary corresponding to of a very small theropod. Turner et al. (2007b) considered *Shanag* as a very basal dromaeosaurid, at a basal polytomy together with microraptorans and velociraptorines, and they noted some features reminiscent to Unenlagiidae. *Shanag* was not included in most ulterior phylogenetic analyses due to its incomplete and poorly informative nature. Nevertheless, the morphology of the putative maxillary fenestra in *Shanag* is clearly different from that of other dromaeosaurids. In *Shanag* this opening differs from that of dromaeosaurids (e.g. *Velociraptor*, *Deinonychus*; Ostrom 1969; Barsbold and Osmolska 1999) in being very reduced and anteroposteriorly short, and being located anteriorly (Fig. 4.1). Moreover, Turner et al. (2007b) indicate that the absence of a promaxillary fenestra was an autapomorphy of *Shanag*, a condition that is considered as diagnostic of Neotheropoda (see Rauhut 2003). In this way, there is some evidence that allow proposing a different interpretation of *Shanag* anatomy. In most paravians the promaxillary fenestra is a slit-like anteriorly located opening (Witmer 1997; Senter et al. 2010), being very similar in shape and position to the structure interpreted by Turner et al. (2007b) as the maxillary fenestra. Moreover, the interpretation of Turner et al. (2007b) of the putatively reduced cranial portion of the antorbital fossa is clearly more reminiscent to the anterior rim of the large maxillary fenestra exhibited by basal Avialae and related taxa (e.g. *Archaeopteryx, Anchiornis, Xiaotingia*; Mayr et al. 2007; Hu et al. 2009; Xu

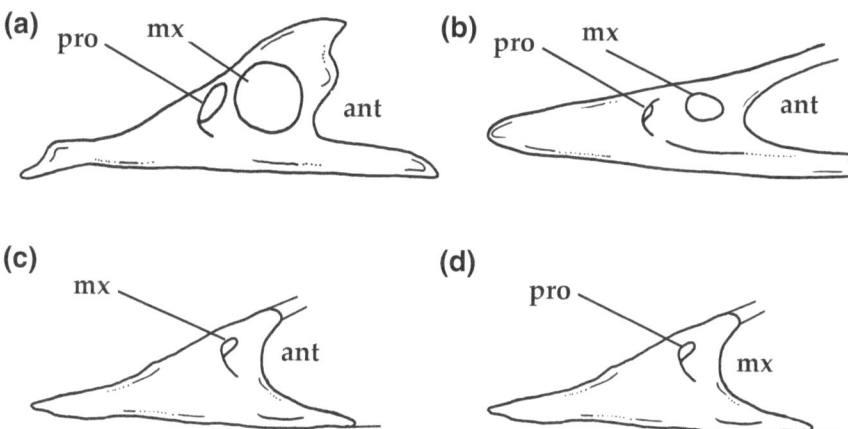

Fig. 4.1 Left maxillae of selected paravians. **a** *Archaeopteryx lithographica*; **b** *Velociraptor mongoliensis*; **c** *Shanag agile* as interpreted by Turner et al. (2007); **d** *Shanag agile* as interpreted here. **a–b** Modified from Senter et al. (2010); **c–d** modified from Turner et al. (2007). Not to scale

et al. 2011). In this way, we reinterpret the antorbital fossa and maxillary fenestra of Turner et al. (2007b) as the anterior rim of the maxillary fenestra and the promaxillary fenestra respectively (Fig. 4.1). *Shanag* was referred to Dromaeosauridae by Turner et al. (2007b) on the basis of large maxillary and dentary teeth, a straight, parallel-sided dentary, and a dorsally displaced maxillary fenestra recessed in a caudodorsally directed depression. However, teeth size of *Shanag* is not different from basal Aves (e.g. *Archaepteryx*; Mayr et al. 2007), troodontids (Sues and Averianov 2008), and microraptorans (Xu 2002), and the morphology of the dentary is clearly more widespread among theropods than previously thought, being widely distributed among basal birds (Zhou et al. 2009; Hu et al. 2010; Agnolín and Novas 2011). Turner et al. (2007b) indicate that *Shanag* resembles Unenlagiidae on the basis of nutrient foramina on external surface of the dentary lying within a deep groove (character 71-1 of Turner et al. 2007b). Regrettably, the presence of such foramina is clearly widespread among paravians, being present in troodontids, basal averaptorans, *Anchiornis*, and *Archaeopteryx* (Agnolín and Novas 2011). On the other hand, *Shanag* shows some features that suggest that this genus may be included within Averaptora, and even among Avialae. *Shanag* exhibits an anteriorly tapering and triangular maxilla, a feature also present in the basalmost avialan *Archaeopteryx* (Turner et al. 2007b) and *Anchiornis* (Hu et al. 2009) and *Xiaotingia* (Xu et al. 2011). In addition, the maxilla of *Shanag* contributes to the narial border and the caudal margin of the narial opening overlaps the rostral border of the antorbital fossa, both features present also in *Archaeopteryx*, but absent in dromaeosaurids (Turner et al. 2007b). Moreover, maxillary teeth show labial longitudinal sulci, a condition seen in microraptorans and unenlagiids (Gianechini et al. 2009; Gianechini and Apesteguía 2011), and anterior maxillary teeth are devoid of serrations, a condition present in microraptorans, unenlagiids, and basal birds (Xu 2002; Agnolín and Novas 2011). Finally, if correctly reinterpreted, the enlarged and rounded maxillary fenestra is a condition which *Shanag* shares with *Anchiornis* and *Archaeopteryx* (Mayr et al. 2007; Hu et al. 2009). In sum, available information suggest *Shanag* as a basal member of Averaptora.

4.5 *Tianyuraptor ostromi*

It was described by Zheng et al. (2009) on the basis of a nearly complete skeleton from the Lower Cretaceous of Liaoning, China. In the strict consensus of the phylogenetic analysis conducted by Zheng et al. (2009) *Tianyuraptor* appears within a polytomy at the base of Dromaeosauridae. Its referral to Dromaeosaridae was based on two features actually present in a wide array of paravians (i.e. dorsal arch of manual ungual I and elongate caudal prezygapophyses and chevrons; Agnolín and Novas 2011). *Tianyuraptor* was considered by Zheng et al. (2009) as a Microraptoria, on the basis of three shared features with members of that clade: laterally sculpted maxilla, shortened manual phalanx III-2, and spatulate pubic symphysis. A sculpted maxilla may constitute a synapomorphy uniting *Tianyuraptor* and Microraptoria, but the remaining features deserve the following comments: the shortened proportions of phalanx III-2 is a condition seen in a wide array of basal avians, including *Archaeopteryx, Jeholornis, Confuciusornis,* and *Anchiornis* (Campbell 2008; Zhou and Zhang 2002; Chiappe et al. 1999; Hu et al. 2009), thus it appears to be more widespread than suggested by Zheng et al. (2009). In regards with the presence of a spatulated pubic boot, it is a condition not only present in microraptorians, but also in basal birds, such as *Anchiornis, Archaeopteryx, Rahonavis, Jeholornis,* and *Confuciusornis* (Forster et al. 1998; Zhou and Zhang 2002, 2003b; Paul 2002; Mayr et al. 2007; Hu et al. 2009). In addition, *Tianyuraptor* lacks several microraptorian apomorphies (see Xu and Wang, 2004), suggesting that this taxon is outside Microraptoria (Zheng et al. 2009). Moreover, as recognized by Zheng et al. (2009) *Tianyuraptor* shows some traits more derived than microraptorans, that are present in unenlagiids and avialans, including an elongate preacetabular process of ilium and strongly concave posterior ischial margin. In sum, we interpret *Tianyuraptor* as an averaptoran of uncertain position.

4.6 *Mahakala omnogova*

It is a small paravian described by Turner et al. (2007a) on the basis of an incomplete skeleton collected in Campanian beds from Mongolia. This minute theropod was summarily described, and in most analyses appears occpying a basal position within Dromaeosauridae, either forming a polytomy with Microraptoria, Unenlagiidae, and with the remaining dromaeosaurids or Eudromaeosauria (Xu et al. 2008; Longrich and Currie 2008), or as the basalmost Dromaeosauridae (Turner et al. 2007a, 2011). Although *Mahakala* is not included in the present analysis, the phylogenetic position inferred by all those authors implies averaptoran affinities for this taxon. Turner et al. (2007a) recognized many features from which *Mahakala* differed from other dromaeosaurids, including a broad ulna, reduced cuppedicus fossa on ilium, vaulted braincase, fibula not contacting distal tarsals, and minute size. Among these traits, a broad ulna, reduced cuppedicus fossa on ilium, and vaulted braincase are features usually regarded as avialan synapomorphies (Senter et al. 2004; Novas 2004), and a broad ulna is also present in microraptorans (Paul 2002). In addition, a distally reduced fibula, lacking contact with distal tarsals is synapomorphic of birds more derived than *Confuciusornis* (i.e. Ornithothoraces). Moreover, *Mahakala* lacks several dromaeosaurid synapomorphies, including elongate prezygapophyses and chevrons on caudal vertebrae (Turner et al. 2007a). In the same way, *Mahakala* resembles derived averaptoran taxa, such as *Buitreraptor* and *Rahonavis* in having a longitudinal groove or ridge near the neurocentral suture of the lateral surface of middle caudal vertebrae, non opisthopubic pelvis, and pubis shorter than femur (Xu et al. 2010). In addition, Turner et al. (2011) reported several features that *Mahakala* shares with averaptorans and avialans, different from the condition seen in dromaeosaurids. Among these features are the presence of small and numerous teeth devoid of carinae and serrations, weakly curved anterior margin of supratemporal fossa, double squamosal articulation for the quadrate, very large foramen magnum, cervical ribs fused to cervical vertebrae, scapula strongly tapering distally, reduced calcaneum, very short and distally located metatarsal I, and elongate pedal phalanges. All these features are present in more derived taxa, including basal birds, such as *Archaeopteryx* and *Rahonavis*, as recognized by Turner et al. (2011). Moreover, *Mahakala* also shows some features that are present in some basal averaptorans and birds, but are absent in dromaeosaurids. As for example, *Mahakala* resembles *Rahonavis* in having caudal prezygapophyses transversely expanded, posterior caudals with a longitudinal lateral ridge (a condition also reported in *Buitreraptor* and *Microraptor*; Hwang et al. 2002; Makovicky et al. 2005), mound-like trochanteric shelf on femur (also present in *Microraptor* and *Buitreraptor*; Hwang et al. 2002; Makovicky et al. 2005), proximally unconstricted metatarsal III, and distal end of metatarsal II without distal flexor pits (see Turner et al. 2011). Moreover, *Mahakala* shares with *Buitreraptor* the everted dorsal margin of the postacetabular blade of ilium, a condition previously thought to be unique of *Buitreraptor* (Gianechini and Apesteguía 2011). In this way, there is an extensive list of features suggesting that *Mahakala* is more nearly related to averaptorans and birds than previously thought.

Mahakala was referred to Dromaeosauridae by Turner et al. (2007a) on the basis of an accessory tympanic recess dorsal to the crista interfenestralis on the braincase, elongate paroccipital processeses with parallel dorsal and ventral margins that twist rostrolaterally distally, and the presence of a ginglymoid distal metatarsal II. However, as detailed by Paul (2002) the morphology of paraoccipital processes in *Archaeopteryx* and dromaeosaurids show a nearly identical morphology (see also Martin 1991). Moreover, the absence of an accessory dorsal tympanic recess is also seen in a wide variety of taxa, including carnosaurs, *Tyrannosaurus,* ornithomimids and troodontids (Turner et al. 2011). Moreover, in most basal averaptorans and birds (with the exception of *Archaeopteryx,* in which this recess is certainly present; Walker 1985; Xu 2002) the absence or presence of such recess cannot be observed due to deficient preservation of specimens. The morphology of disal metatarsal II is more widespread than previously thought, and its distribution is equivocal among Paraves (Agnolín and Novas 2011). More recently, Turner et al. (2011) added new characters that prompted the inclusion of *Mahakala* within Dromaeosauridae: anterior cervical centrum extends beyond the posterior limit of the neural arch, stalk-like parapophyses on dorsal vertebrae, and anterior tympanic recess anteriorly located. However, the extension of the cervical centrum with respect to the neural arch is a feature that appears to be very variable along the cervical vertebrae of paravians, and was dismissed as a dromaeosaurid synapomorphy by Agnolín and Novas (2011). In *Mahakala* the stalk-like parapophyses of dorsal vertebrae differs from that of dromaeosaurids on the extremely short pedicel (Turner et al. 2011), resembling in this way, *Confuciusornis* and more derived birds (Agnolín and Novas 2011). Moreover, pedunculated parapophyses were considered by Agnolín and Novas (2011) as diagnostic of more inclusive clades, probably Maniraptora. In this way, the only feature that stands as a probable synapomorphy uniting *Mahakala* with dromaeosaurids is the anteriorly placed anterior tympanic recess (Turner et al. 2011). However, it must be noted that tympanic information is not available for several basal averaptorans, including Microraptoria, Unenlagiidae, *Anchiornis* and *Xiaotingia,* as well as, most basal birds. This tends to blur the synapomorphic condition of such anatomical trait.

Concluding, the absence of clear dromaeosaurid synapomorphies in join with the presence of derived averaptoran and avialan traits (e.g. reduced cuppedicus fossa on ilium, distally reduced fibula) support averaptoran, and even avialan affinities for *Mahakala.*

4.7 *Jinfengopteryx elegans*

This taxon was described on the basis of a nearly complete specimen from the Lower Cretaceous of China (Ji et al. 2005). *Jinfengopteryx* was interpreted as a basal avialan, probably related with *Archaeopteryx* (Ji et al. 2007; Yuan 2008), as coming from the Lower Cretaceous of China. This theropod was lately considered by Xu and Norell (2004) as belonging to the Troodontidae, based mainly on skeletal proportions and tooth morphology. More recently, Turner et al. (2007a) reinforced the troodontid affiliation of *Jinfengopteryx* on the basis of an extensive phylogenetic analysis. These authors, in an Adam's consensus tree obtained from 1296 MPTs resulted in a single tree with a nearly fully resolved topology that nested *Jinfengopteryx* within Troodontidae. However, if a Strict Consensus is applied on the analysis of Turner et al. (2007a) *Jinfengopteryx* s excluded from

Troodontidae, but results part of a large polytomy within Paraves. In spite of such methodological incongruences, we will analyze the putative troodontid synapomorphies shared by *Jinfengopteryx* with in remaining troodontids. Turner et al. (2007a) indicated several traits (their characters 21-1, 48-1, 51-1, 70-1, 71-1, 85-1, 89-1, 127-1, 203-1, 208-1, 224-1, 225-1, 229-1) as diagnostic of Troodontidae. But, characters 48-1, 51-1, 85-1, 127-1, 224-1, 225-1 and 229-1 are not observable in the holotype of *Jinfengopteryx*. Character 21-1 of Turner et al. (2007a) consists in the presence of a flat internarial bar. This condition, however is not only present in troodontids, but also in *Anchiornis* and *Archaeopteryx*, being unknown in unenlagiid specimens. Thus, its phylogenetic significance is up to now uncertain. The presence of a subtriangular dentary (character 70-1), a row of nutrient foramina that lie within a deep groove at lateral face of dentary (character 71-1), and anterior dentary teeth closely appressed (character 89-1) were considered as widespread among paravians, and were not recovered as diagnostic of any paravian clade (Agnolín and Novas, 2011). The presence of a subarctometatarsalian pes (character 203-1) is currently considered diagnostic of the node Paraves (Agnolín and Novas 2011). Finally, another trait considered by Turner et al. (2007a) as a diagnostic trait of Troodontidae not shared with *Jinfengopteryx* is the asymmetrical foot (character 208-1). However, the foot of *Jinfengopteryx* is very poorly preserved, and only the proximal end of metatarsals has been preserved, being thus the condition of the foot in *Jinfengopteryx* remains uncertain. Besides *Jinfengopteryx* shows derived averaptoran traits, including minute body size, elongate forelimbs (Turner et al. 2011), very elongate metacarpal I, short and unspecialized pedal digit II, short ischial peduncle of ischium, and thin ischial shaft (Ji et al. 2005; Ji and Ji 2007; Yuan 2008; see above). However, *Jinfengopteryx* shows some plesiomorphic traits when compared with other averaptorans, including a distally expanded scapula and short forelimbs. In this way, due to the equivocal skeletal features exhibited by *Jinfengopteryx*, this theropod is here excluded from Troodontidae and it is interpreted as Averaptora *incertae sedis*.

4.8 *Unquillosaurus ceiballi*

It was described by Powell (1979) on the basis of a large pubis of an indeterminate carnosaurian theropod coming from the Late Cretaceous of NW Argentina. Latter, *Unquillosaurus* was considered as a derived paravian by Novas and Agnolín (2004), and as a dromaeosaurid (Norell and Makovicky 2004). However, *Unquillosaurus* lacks any apomorphic feature that may allow its referral to Dromaeosauridae. Recently, Carrano et al. (2012) considered that *Unquillosaurus* belongs to Carcharodontosauridae. They indicated that the distal end of the pubis is strongly abraded, and when complete may have an anteroposteriorly extended distal pubic boot. However, although abraded, the anterior margin of the pubis of Unquillosaurus clearly indicates that the pubis lacks its anterior projection. Moreover, the distally thin pubic boot is a condition that allows inclusion of

Unquillosaurus within Coelurosauria, clearly differing from the morphology seen in Carcharodontosauridae (Benson et al. 2010). On the contrary, *Unquillosaurus* resembles averaptorans in having a reduced pubic symphysis and an anteroposteriorly short pubic boot, without well developed anterior projection, a condition shared with microraptorans and basal birds as *Archaeopteryx* and *Jeholornis*. Moreover, *Unquillosaurus* shows a reduced ischial process of pubis, a condition reminiscent of Unenlagiidae and Avialae (see above). In this way, it is probable that *Unquillosaurus* may represents a very large member of the Averaptora.

4.9 *Pamparaptor micros*

This minute theropod was described as an unenlagiid by Porfiri and collaborators (2011). The specimen consist on an incomplete foot of a small theropod dinosaur. Porfiri et al. (2011) interpreted *Pamparaptor* as an unenlagiid because it shares several common features with *Neuquenraptor* (considered here as a junior synonym of *Unenlagia*), including subarctometatarsal metatarsus, metatarsal IV with a posterolateral flange, proximal half of metatarsal III with an extensor sulcus, and metatarsal II with a lateral expansion over the caudal surface of metatarsal III. However, as pointed out by Agnolín and Novas (2011; see above) all these features are probably more widespread than previously thought, and their status as unenlagiid or dromaeosaurid synapomorphies are discussable. In this way, we consider *Pamparaptor* as an Averaptora *incertae sedis*, until more complete and detailed analysis of the specimen became available.

4.10 European Dromaeosaurids

Makovicky et al. (2005; SI) suggested that some European dromaeosaurids may be included within Unenlagiidae. However, the scanty material belonging to European dromaeosaurids does not particularly resemble those of derived averaptorans, including unenlagiids. For example, *Pyroraptor olympus* Allain and Taquet (2000), from the Latest Cretaceous of France retains a plesiomorphically short ulna, and the phalanx 2-II lacks the weak heel seen in unenlagiids and basal birds, exhibiting a well developed and symmetrical structure comparable to that seen in typical dromaeosaurids (Longrich and Currie 2008; Agnolín and Novas 2011). The genus *Variraptor mechinorum* Le Loeuff and Buffetaut (1998) from the Upper Cretaceous of France, shows a plesiomorphic sacrum with only five coosified vertebrae, cervical vertebrae with very large epipophyses and a robust and stout humerus with a cranially oriented deltopectoral crest, a combination of plesiomorphic traits not seen in any known averaptoran. Moreover, all known dromaeosaurid-like teeth recovered from several fossil localities of Jurassic and Cretaceous ages in Europe (e.g. Buffetaut et al. 1986; Canudo et al. 1997; Zinke 1998; Rauhut

2002) lack the synapomorphic traits seen in unenlagiids, such as absence of serrations in anterior and posterior carinae, and presence of longitudinal sulci along the teeth crowns (Ezcurra 2008; Gianechini et al. 2009; Gianechini and Apesteguía 2011). Concluding, most (if not all) remains of European dromaeosaurid-like specimens lack derived traits that may unite them with Unenlagiidae among averaptoran theropods.

References

Agnolín FL, Novas FE (2011) Unenlagiid theropods: are they members of Dromaeosauridae (theropoda, Maniraptora). An Acad Bras Ciênc 83:117–162

Allain R, Taquet P (2000) A new genus of Dromaeosauridae (Dinosauria, Theropoda) from the Upper Cretaceous of France. J Vert Paleont 20:404–407

Averianov AO, Sues HD (2007) A new troodontid (Dinosauria: Theropoda) from the Cenomanian of Uzbekistan, with a review of troodontid records from the territories of the former Soviet Union. J Vert Paleontol 27: 87–98.

Barsbold R, Osmólska H (1999) The skull of *Velociraptor* (Theropoda) from the Late Cretaceous of Mongolia. Acta Palaeont Pol 44:189–219

Buffetaut E, Marandat B, Sigé B (1986) Découverte de dents de Deinonychosaures (Saurischia, Theropoda) dans le Crétacé supérieur du sud de la France. CR Acad Sci Paris 303:1393–1396

Burnham D (2008) A review of the early Cretaceous Jehol group and a new paradigm for the origin of flight. Oryctos 7:27–42

Campbell KE (2008) The manus of archaeopterygians: implications for avian ancestry. Oryctos 7:13–26

Canudo JI, Amo O, Cuenca-Bescós G, Meléndez A, Ruiz-Omeñaca JI, Soria AR (1997) Los vertebrados del Tithónico-Barremiense de Galve (Teruel, España). Cuad Geol. Iber 23:209–241

Chiappe LM, Ji S, Ji Q, Norell MA (1999) Anatomy and systematics of Confuciusornithidae from the Late Mesozoic of northeastern China. Bull Amer Mus Nat Hist 242:1–89

Currie PJ (2000) Theropods from the Cretaceous of Mongolia. Cambridge University Press, The Age of Dinosaurs in Russia and Mongolia. Cambridge, pp 434–455

Ezcurra MD (2008) Theropod remains from the latest Cretaceous of Colombia and their implications on the palaeozoogeography of western Gondwana. Cret Res 30: 1339–1344.

Forster C, Sampson S, Chiappe LM, Krause D (1998) The theropod ancestry of birds: new evidence from the Late Cretaceous of Madagascar. Science 279:1915–1919

Gianechini FA, Apesteguía S (2011) Unenlagiinae revisited: dromaeosaurid theropods from South America. An Acad Bras Ciênc 83:163–197

Gianechini FA, Apesteguía S, Makovicky PJ (2009) The unusual dentition of *Buitreraptor gonzalezorum* (Theropoda, Dromaeosauridae), from Patagonia, Argentina: new insights on the unenlagine teeth. Ameghiniana 52:36A

Hu D, Hou L, Zhang L, Xu X (2009) A pre-*Archaeopteryx* troodontid theropod from China with long feathers on the metatarsus. Nature 461:640–643

Hu D, Li L, Hou L, Xu X (2010) A new sapeornithid bird from China and its implications for early avian evolution. Act Geol Sin 84:472–482

Hwang SH, Norell MA, Qiang J, Keqin G (2002) New specimens of *Microraptor zhaoianus* (Theropoda: Dromaeosauridae) from Northeastern China. Amer Mus Novit 3381:1–44

Ji S, Ji Q (2007) *Jinfengopteryx* compared to *Archaeopteryx*, with comments on the mosaic evolution of long-tailed avialan birds. Act Geol Sin 81:337–343

Ji Q, Ji S, Lu J, You H, Chen W, Liu Y, Liu Y (2005) First avialan bird from China (*Jinfengopteryx elegans* gen. et sp. nov.). Geol Bull China 24:197–205

Le Loeuff J, Buffetaut E (1998) A new dromaeosaurid theropod from the Upper Cretaceous of southern France. Oryctos 1:105–112.

Longrich NR, Currie PJ (2008) A microraptorine (Dinosauria-Dromaeosauridae) from the Late Cretaceous of North America. PNAS 106:5002–5007

Lü J-C, Xu L, Zhang X-L, Ji Q, Jia S-H, Hu W-Y, Zhang J-M, Wu Y-H (2007) New dromaeosaurid dinosaur from the Late Cretaceous Qiupa Formation of Luanchuan area, western Henan, China. Geol Bull China 26:777–786

Makovicky PJ, Apesteguía S, Agnolín FL (2005) The earliest dromaeosaurid theropod from South America. Nature 437:1007–1011

Martin LD (1991) Mesozoic birds and the origin of birds. In: Schultze HP, Trueb L (eds) Origin of the higher groups of tetrapods. Comstock, Ithaca, N.Y., pp 485–540

Mayr G, Pohl B, Hartman S, Peters DS (2007) The tenth skeletal specimen of *Archaeopteryx*. Zool Jour Lin Soc 149:97–116

Norell MA, Makovicky PJ (2004) Dromaeosauridae. In: Weishampel DB, Dodson P, Osmolska H (eds) The Dinosauria (2nd edn). University of California Press, Berkeley, pp 196–209

Novas FE (2004) Avian traits in the ilium of *Unenlagia comahuensis* (Maniraptora, Avialae). In: Currie PJ, Koppelhus EB, Shugar MA, Wright JL (eds) Feathered dragons: studies on the transition from dinosaurs to birds. Indiana University Press, Bloomington, pp 150–166

Novas FE, Agnolín FL (2004) *Unquillosaurus ceibali* Powell, a giant maniraptoran (Dinosauria, Theropoda) from the Late Cretaceous of Argentina. Rev Mus Arg Cienc Nat 6:61–66

Osmólska H (1982) *Hulsanpes perlei* n. g. n. sp. (Deinonychosauria, Saurischia, Dinosauria) from the Upper Cretaceous Barun Goyot formation of Mongolia. N Jahr Geol Palaeont 1982:440–448

Ostrom JH (1969) Osteology of *Deinonychus antirrhopus*, an unusual theropod from the Lower Cretaceous of Montana. Bull Peabody Mus Nat Hist 30:1–165

Paul GS (2002) Dinosaurs of the Air. The John Hopkins University Press

Porfiri JD, Calvo JO, Dos Santos D (2011) A new small deinonychosaur (Dinosauria: Theropoda) from the Late Cretaceous of Patagonia, Argentina. An Acad Bras Ciênc 83:109–116

Powell JE (1979) Sobre una asociación de dinosaurios y otras evidencias de vertebrados del Cretácico Superior de la región de la Candelaria, Prov. de Salta, Argentina. Ameghiniana 16:191–204

Rauhut OWM (2002) Dinosaur teeth from the Barremian of Una, Province of Cuenca, Spain. Cret Res 23:255–263

Rauhut OWM (2003) The interrelationships and evolution of basal theropod dinosaurs. Spec Pap Palaeont 69:1–213

Senter P, Barsbold R, Britt B, Burnham D (2004) Systematics and evolution of Dromaeosauridae (Dinosauria: Theropoda). Bull Gunma Mus Nat Hist 8:1–20

Senter P, Kirkland JI, Bird J, Bartlett JA (2010) A new troodontid theropod dinosaur from the Lower Cretaceous of Utah. PLoS ONE 5:e14329

Turner AH, Pol D, Clarke JA, Erickson GM, Norell MA (2007a) A basal dromaeosaurid and size evolution preceding avian flight. Science 317:1378–1381

Turner AH, Hwang SH, Norell MA (2007b) A small derived theropod from Oösh, Early Cretaceous, Baykhangor Mongolia. Amer Mus Novit 3557:1–27

Turner AH, Makovicky PJ, Norel MA (2007c) Feather Quill Knobs in the Dinosaur *Velociraptor*. Science 21317:1721

Turner AH, Pol D, Norell MA (2011) Anatomy of *Mahakala omnogovae* (Theropoda: Dromaeosauridae), Tögrögiin Shiree, Mongolia. American Museum Novitates 3722:1–66

Walker AD (1985) The braincase of *Archaeopteryx*. In: Hecht MK, Ostrom JH, Viohl G, Wellnhofer P (eds) The beginnings of birds: proceedings of the international *Archaeopteryx* conference, Eichstätt, 1985. Freunde des Jura-Museum Eichstätt, Eichstätt, pp 123–134

Witmer LM (1997) The evolution of the antorbital cavity of archosaurs: a study in soft-tissue reconstruction in the fossil record with an analysis of the function of pneumaticity. Soc Vert Palaeont Mem 3:1–73

Xu X (2002) Deinonychosaurian fossils from the Jehol Group of western Liaoning and the coelurosaurian evolution. Dissertation for the doctoral degree. Chinese Academy of Sciences, Beijing

Xu X, Norell MK (2004) A new troodontid dinosaur from China with avian-like sleeping posture. Nature 431:838–841

Xu X, Wang X-L (2004) A new Dromaeosaur (Dinosauria: Theropoda) from the Early Cretaceous Yixian Formation of Western Liaoning. Vert Palas 42:11–119

Xu X, Zhang F (2005) A new maniraptoran dinosaur from China with long feathers on the metatarsus. Naturwissenschaften 92:173–177

Xu X, Zhao Q, Norell MA, Sullivan C, Hone D, Erickson PG, Wang X, Han F, Guo Y (2008) A new feathered dinosaur fossil that fills a morphological gap in avian origin. Chinese Sci Bull 54:430–435

Xu X, Choinere J, Pittman M, Tan Q, Xiao D, Li Z, Tan L, Clark J, Norell M, Hone DWE, Sullivan C (2010) A new dromaeosaurid (Dinosauria: Theropoda) from the Upper Cretaceous Wulansuhai Formation of Inner Mongolia, China. Zootaxa 2403:1–9

Xu X, You H, Du K, Han F (2011) An *Archaeopteryx*-like theropod from China and the origin of Avialae. Nature 475:465–470

Yuan C (2008) A new genus and species of Sapeornithidae from Lower Cretaceous in Western Liaoning, China. Acta Geol Sinica 82:48–55

Zheng X, Xu X, You H, Zhao Q, Dong Z (2009) A short-armed dromaeosaurid from the Jehol Group of China with implications for early dromaeosaurid evolution. Proc Royal Soc London B 277:211

Zhou Z, Clarke J, Zhang F (2009) Insight into diversity, body size and morphological evolution from the largest Early Cretaceous enantiornithine bird. J Anat 212: 565–577.

Zhou ZH, Zhang FC (2002) A long-tailed, seed-eating bird from the Early Cretaceous of China. Nature 418:405–409

Zhou ZH, Zhang FC (2003a) Anatomy and systematics of the primitive bird *Sapeornis chaoyangensis* from the Early Cretaceous of Liaoning, China. Canadian J Earth Sci 40:731–747

Zhou ZH, Zhang FC (2003b) *Jeholornis* compared to *Archaeopteryx*, with a new understanding of the earliest avian evolution. Naturwissenschaften 90:220–225

Zinke J (1998) Small theropod teeth from the Upper Jurassic coal mine of Guimarota (Portugal). Palaont Zeitsch 72:179–189

Chapter 5
Discussion

Present analysis invites to review previous hypotheses regarding the acquisition of evolutionary novelties towards the line to birds, especially those of appendicular skeleton and integumentary structures.

5.1 Evolution of Feathers and Wings Among Basal Paravians

Recently Hu et al. (2009) and Witmer (2009) interpreted that the four-winged condition evolved in the common paravian ancestor, because these authors accepted *Microraptor*, *Anchiornis*, and *Pedopenna* as basal members of Dromaeosauridae, Troodontidae and Avialae, respectively (Fig. 5.1). However, in the context of the phylogeny defended here, the development of hindlimb wings (producing a four-winged pattern) did not occur at the base of Paraves, but at the base of Averaptora: *Microraptor* exhibits well-developed hindlimb wings, with 14 pennaceous feathers attached to the metatarsals (Xu et al. 2003), a condition that is also seen in *Anchiornis* (Hu et al. 2009), as well as in the basal avialan *Pedopenna* (Xu and Zhang 2005). The presence of very long feathers in the femur of *Xiaotingia* also suggests a tetrapterygian condition for this genus (Xu et al. 2011a). However, a trend towards the reduction of hindwings occurred among avialans: in *Anchiornis*, although the hindwings are extensive, the metatarsal feathers have symmetrical vanes, condition that is usually considered as indicative of non-aerodynamical functions (Feduccia 1996). In the basal avialan *Pedopenna*, the vanes of distal metatarsal feathers are also symmetrical, but they are proportionally smaller and weaker than in *Microraptor* and *Anchiornis* (Xu and Zhang 2005). In *Archaeopteryx*, the metatarsal feathers are still present, although strongly reduced and not attached to the metatarsus (Christiansen and Bonde 2004; Hone et al. 2010). Leg feathers have been also confirmed in *Confuciusornis*, *Longipteryx* and other enantiornithes (Zhang and Zhou 2004), although they are extremely reduced in size. In non-enantiornithine ornithothoracine birds the leg feathers are totally absent (Zhou and Zhang 2006). Instead, they are replaced by pedal scales, which may be secondarily derived structures diagnostic of Ornithothoraces (Hu

F. L. Agnolín and F. E. Novas, *Avian Ancestors,* SpringerBriefs in Earth System Sciences,   49
DOI: 10.1007/978-94-007-5637-3_5, © The Author(s) 2013

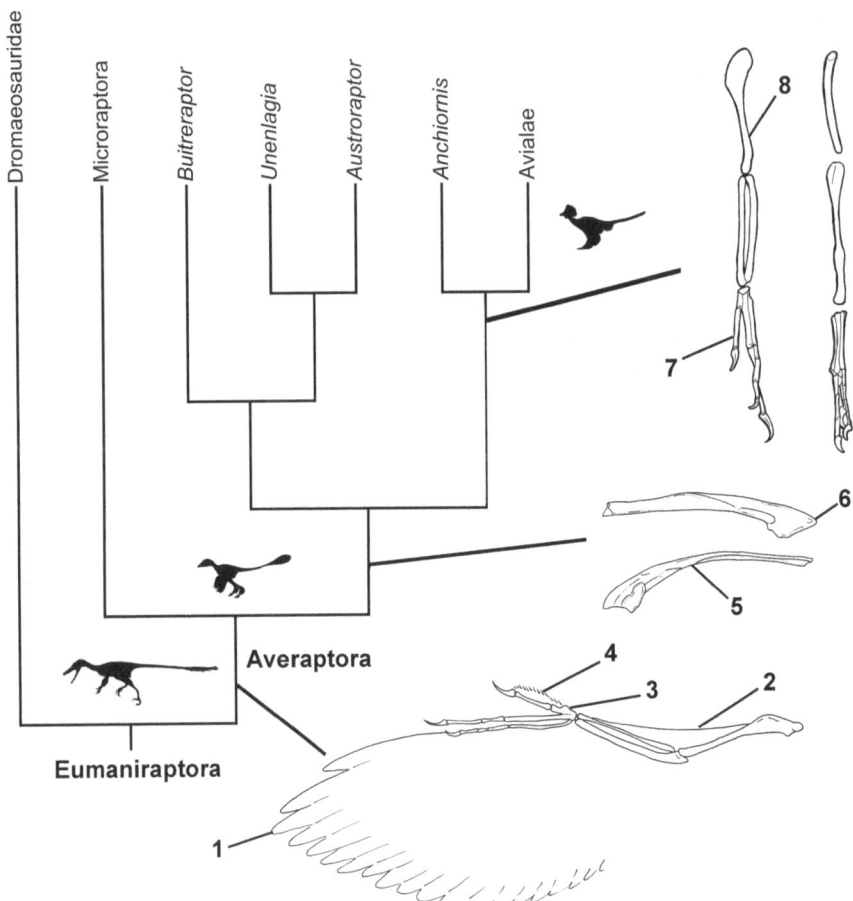

Fig. 5.1 Diagram of eumaniraptoran phylogenetic relationships showing main anatomical adquisitions related to flight. Forewing of *Microraptor gui* showing main characters allowing a powewerful flight: *1* asymmetrical remiges; *2* propatagial tendons; *3* expanded extensor process on metacarpal I; *4* alula. Scapula of *Unenlagia comahuensis*; *5* twisted scapular blade; *6* subtriangular and pointed acromion. Left fore and hindlimb of *Archaeopteryx lithographica*; *7* extremely elongate manus; *8* very robust and elongate humerus (much more robust and larger than femur)

et al. 2009). In sum, available information indicates that hindwings were progresively reduced and lost in the line to birds, as previously advocated by Xu and Zhang (2005).

In relation with the previous topic, a similar trend in the reduction in the number of secondary remiges manifest in the forelimbs: it is 12 in basal maniraptorans (e.g., the oviraptorosaurian *Similicaudipteryx*; Xu et al. 2010), a number that is similar to that in *Anchiornis* and *Archaeopteryx* for which 12 or probably 13 secondary remiges have been reported (Hu et al. 2009; Christiansen and Bonde 2004). In more derived birds, the number of remiges is even lower, with 10 remiges proposed for *Rahonavis* (Forster et al. 1998), and less than 10 remiges in Enantiornithes (e.g., 8 in *Eoalulavis*; Sanz et al. 1996). In modern birds the number of remiges is highly variable, although the presence of 10 secondary remiges appears to be the more widespread condition among living birds (Feduccia 1996).

However, certain dromaeosaurids (e.g., *Velociraptor*) and *Microraptor* seem to depart from this trend towards the reduction in the number of remiges: 14 secondary remiges are attached to the ulna in *Velociraptor* (Turner et al. 2007b), and in *Microraptor* at least 18 of these feathers have been counted (Chatterjee and Templin 2007). Taken alone the number of secondary remiges, the presence of more than 14 of these feathers seem to represent a derived condition shared by *Microraptor* and *Velociraptor*, but this alternative needs to be confirmed with additional specimens preserving the feather covering.

A puzzling situation for microraptorians is that two different kinds of feathers are present in the best known members of this group: simple, branched feather-like structures are documented in the 150 cm long *Sinornithosaurus milleni*, and considerably more complex, pennaceous feathers exhibited by *Microraptor gui*, which surpass 77 cm in length (Xu et al. 1999, 2000, 2003; Xu and Guo 2009). Character presence or absence of remiges has been coded in the present matrix, but the lack of information for most of the studied taxa precludes recognition of a pattern. Moreover, recent analysis indicate that the two-dimensional preservation of specimens during fossilization makes the identification of different kind of feather difficult due to overlapping feather structures in vivo (Foth 2011).

In regards with the Scansoriopterygidae, Zhang et al. (2008) have explained the absence of pennaceous feathers in these coelurosaurs as a consequence of a secondary loss of flight capabilities. However, in the present phylogeny, Scansoriopterygidae are located far from Avialae, and the absence of modern-like feathers is better interpreted as primary rather than to the result of a secondary reversal.

5.2   Osteological and Integumental Modifications Related to the Origin of Flight

Novas and Puerta (1997) suggested that most relevant differences in the line of theropods to birds have to do with changes in skeletal proportions (see also Xu et al. 2011a). In this regard, *Anchiornis* exhibits forelimb proportions that are more derived than those of *Microraptor*, thus closely resembling the condition present in basal birds (e.g., *Archaeopteryx*, *Jeholornis*) (Fig. 5.1). For example, in *Anchiornis* the humerus is longer than the femur, and its transverse with equals that of the latter bone; forelimb lenght in *Anchiornis* is 80 % of hindlimb length, and its elongate hand represents about 130 % of femoral length (Xu et al. 2008). This set of modifications in the forelimbs may reflect improved flying capabilities in avialans.

Several other features related with the acquisition of flying control (e.g., alula, protopatagium, automatic control of forearms, development of muscles that flex the forearms) have developed early in the evolution of averaptorans, and they are reviewed in light of the present phylogeny (Fig. 5.1).

The propatagium is an integumentary structure that fills the space in front of the flexed wing, being considered as a very important condition for flight control (Paul 2002); in addition, the ligamentum propatagiale may also support the distal portion of the wing against drag (Vasquez 1994). Several authors (see Paul 2002) proposed that the propatagium constituted a diagnostic trait of derived birds (i.e. Ornithurae). Although the fossil record of propatagium is still patchy, presence of this soft structure has been recently documented in *Microraptor* (see Xu et al. 2003), *Anchiornis* (see Hu et al. 2009), *Archaeopteryx* (Martin and Lim 2005), and Enantiornithes (e.g., *Noguerornis*; Chiappe and Lacasa Ruiz 2002). Moreover, all known averaptorans (e.g. *Sinornithosaurus, Anchiornis, Jeholornis, Confuciusornis*) show a well-developed extensor process on the carpometacarpus (Paul 2002), an osteological correlate of the insertion site of propatagial tendons (Vasquez 1994). Presence of extensor process on carpometacarpus may indicate that most averaptorans possesed a well-developed propatagium, and that this modern wing design and control already evolved in the common ancestor of Averaptora (Fig. 5.1).

As Vasquez (1992, 1994) pointed out, the modern avian wrist possesses the ability to synchronize flexion of extension of the elbow and wrist joints automatically. This kind of automatic mechanism of the wing is widely accepted (see Vasquez 1994) as an indispensable requirement for the powered and well-controlled flight seen in all modern birds. Vasquez (1994) indicates two main osteological features as indicative of automatic wing coordination: the presence of a groove at the distal-dorsal surface of ulna, and a well developed extensor process on metacarpal I for the insertion of the M. extensor metacarpi radialis (Campbell 2008). The presence of a relatively well-developed extensor process on metacarpal I is corroborated in *Sinornithosaurus, Microraptor, Anchiornis,* and *Archaeopteryx* (Paul 2002; Campbell 2008; Xu et al. 2008), suggesting that the presence of an automatic mechanism for flight may be traced back to the base of Averaptora (Fig. 5.1).

Other modifications documented in early averaptorans regards with the development and orientation of the acrocoracoidal process. This process (also named "biceps tubercle" in non-avian theropods; Ostrom 1976) serves as site for insertion of the M. biceps brachii, the chief flexor of the avian forearm (Sereno 2004; Jasinoski et al. 2006). In basal eumaniraptorans (e.g., *Deinonychus, Bambiraptor, Sinornithoides*; Ostrom 1969; Burnham et al. 2000; Currie and Zhiming 2001) the acrocoracoidal process is small and rounded. In basal averaptorans (*Sinornithosaurus, Microraptor*; Xu 2002; Xu et al. 2003), however, the acrocoracoid process is bigger and more laterally projected, being caudally connected with a sharp and acute ridge that runs along the lateral coracoidal surface (Xu 2002) (Fig. 5.2). The unenlagiid *Buitreraptor* shows a larger and craniolaterally projected acrocoracoidal process, conditions that *Buitreraptor* shares with *Archaeopteryx* and more derived birds (Paul 2002; Makovicky et al. 2005).

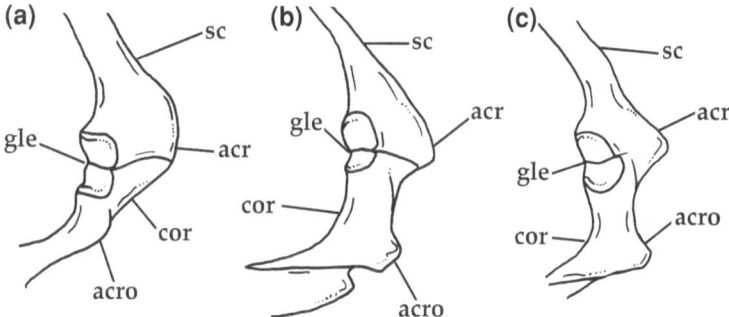

Fig. 5.2 Scapulocoracoids of selected paravian theropods in right lateral view. **a** *Bambiraptor feinbergi*. **b** *Buitreraptor gonzalezorum*. **c** *Archaeopteryx lithographica*. **a**, **c**, modified from Paul (2002); **b** modified from Makovicky et al. (2005). Not to scale

In sum, averaptorans shows the acquisition of two functionally significant sites for muscle attachment (i.e., the acrocoracoidal process and the extensor process on metacarpal I) which are present in the avian wing, suggesting that the ability of flapping flight was already developed at the base of Averaptora.

The set of modifications described above in averaptorans are in concert with a remodeling in the pectoral girdle, which had important consequences in the arc of motion of the forelimbs. The presence of a laterodorsally oriented glenoid cavity on the scapulocoracoid in derived birds (i.e. Ornithurae; Senter 2006) is considered as an unambiguous prerequisite for flapping flight (Fig. 5.1). Novas and Puerta (1997) indicated that the external surface of the scapular blade and the glenoid in *Unenlagia* was laterodorsally exposed, a condition resembling derived flying birds. Such reconstruction was critizised by some authors (Carpenter 2002; Senter 2006) who suggested that the glenoid in *Unenlagia* probably faced posteroventrally, as plesiomorphically occurs in Theropoda. However, in *Unenlegia* the glenoid surface curves in such as a way that its floor tends to be dorsolaterally oriented, implying that the continuation of the glenoid surface into the coracoid must also be dorsally faced, a condition also seen in *Buitreraptor* (Novas 2009). In this way, the scapulae of unenlagiids lie close to the vertebral column, dorsal to the ribcage, with the flat costal surface of the scapular blade facing ventrally, a condition seen in microraptorans (i.e. *Microraptor*), basal avialans (e.g. *Archaeopteryx, Rahonavis*), and ornithothoracine birds (Senter 2006), in which the shoulder socket sits high on the back, and the margins of the glenoid are smooth, thus this surface becomes shalower and consequently more continuous with the rest of the lateral surface of scapula (Burnham 2008). In sum, the lateral orientation of the scapular glenoid in unenlagiids (and probably also in other basal averaptorans), together with the absence of acute ridges delimiting the glenoid cavity, suggest that the humerus in these taxa was able to be elevated close to the vertical plane, as proposed by Novas and Puerta (1997) (Figs. 5.1, 5.2).

It is important to mention that scansoriopterygids retained a caudoventrally oriented glenoid, a subrectangular coracoid with reduced biceps tubercle, and a distally fan-shaped scapular blade, all representing plesiomorphic character states in respect to paravians.

Available information indicates that theropods acquired the ability to fly at the base Averaptora. At this node, main osteological characters correlated to flapping flight can be recognized, as well as integumentary modifications (e.g., alula, asymmetric feathers in ulna and manus, propatagium; Xu et al. 2003; Zhang 2004; Senter et al. 2004).

5.3 Body Size Increase and Loss of Flying Capabilities Among Paravians

Several authors (e.g., Paul 2002; Turner et al. 2007a, b; Senter 2007) interpreted that the common paravian ancestor was a small-sized, flying animal, and that flying capabilities were independently lost in different deinonychosaurian lineages, in association with an increase in body mass. Turner et al. (2007a; see also Turner et al. 2011), for example, hypothesized that Dromaeosauridae underwent three parallel trends in body size increase: one corresponding to *Deinonychus*, another one to *Unenlagia*, and a third one to the clade formed by *Utahraptor* and *Achillobator*. In this evolutionary context, Turner et al. (2007b) considered that aerodynamical capabilities became lost in large-bodied dromaeosaurids, which reduced their set of forelimb feathers (as suggested by the poor development, or absence, of papilae for feather attachment along the caudal margin of ulna). However, in the context of the phylogeny here defended, a single event of body size increase is recovered among dromaeosaurids, and it corresponds to the node made up by *Deinonychus* + (*Atrociraptor* + (*Utahraptor* + *Achillobator*)). Outside Dromaeosauridae, an increase in body size is also manifested in Unenlagiidae, with the basal and turkey-sized *Buitreraptor*, as sister taxon of the ostrich-sized *Unenlagia* and the large *Austroraptor*, reaching approximately 5 m long. Probably, secondary remiges were already present in basal paravians or eumaniraptorans (or more inclusively in tetanurines, as suggested by the presence of feather ulnar papilae in the early carcharodontosaurid *Concavenator*; Ortega et al. 2010), but development of large, asymmetrical secondary remiges for aerodynamic purposes, apparently occurred at the base of Averaptora, with the acquisition of flying cappabilities.

5.4 Independent Origin of Flying Capabilities Among Paravians

Senter (2007) noted the existence of several derived traits shared by microraptorians and unenlagiids (*Rahonavis*, in particular), but according to the results of his phylogenetic analysis, he explained these birdlike characteristics as acquired within each dromaeosaurid clade independently from Aves and also independently from each other. In agreement with this view, Zheng et al. (2009) recently described the short-armed basal dromaeosaurid *Tianyuraptor* as a possible early microraptorine, indicative that more derived, long-armed members of this clade might have developed aerial capabilities independently from birds (Xu et al. 2003; Chatterjee and Templin 2007). The present study, however, removes Microraptoria from Dromaeosauridae, *Tianyuraptor* from Microraptoria, and *Rahonavis* from Unenlegiidae. As a result, our analysis supports that the acquisition of aerial locomotion is more parsimoniously recovered as occurred just only once among paravians, that is at the common ancestor of Averaptora.

5.5 Averaptoran Radiation and Center of Origin of Birds

Available chronological and phylogenetic information strongly suggests that for Middle Jurassic times, at least, main maniraptoran theropods, including Avialae, were already diversified. *Anchiornis*, here posited as the sister-group of Avialae, is at least 150 Ma old, thus implying at least a Lower Jurassic origin for most maniraptoran theropods, including Avialae (Chatterjee 1999). It must be said that some molecular clock analyses considered a probable Triassic origin for birds (Kumar and Hedges 1998).

Xu and Zhang (2005) proposed a Laurasian, or more precisely Asiatic, origin for birds, indicating that most plesiomorphic representatives of Troodontidae and Dromaeosauridae were recorded from Eastern Asia. However, in the present phylogeny, Unenlagiidae is depicted as the immediate sister-group of *Anchiornis* + Avialae, and all presently known unenlagiids came from South America (Makovicky et al. 2005; Novas 2009). In addition, the oldest known Avialae, *Archaeopteryx*, has been recorded in Europe, and the basal bird *Rahonavis* comes from another Gondwanan landmass, Madagascar. In view of the extremely incomplete fossil record of basal paravians, is not possible to confidently establish a center of origin for the Avialae, mainly considering that their diversification occurred during the Mid-Jurassic, when continents were joined in a single landmass.

References

Burnham D (2008) A review of the Early Cretaceous Jehol group and a new paradigm for the origin of flight. Oryctos 7:27–42

Burnham D, Currie PA, Bakker R, Zhou Z, Ostrom J (2000) Remarkable new birdlike dinosaur (Theropoda: Maniraptora) from the Upper Cretaceous of Montana. Univ Kansas Paleont Contrib 13:1–14

Campbell KE (2008) The manus of archaeopterygians: implications for avian ancestry. Oryctos 7:13–26

Carpenter K (2002) Forelimb biomechanics of nonavian theropod dinosaurs in predation. Senck Leth 82:59–76

Chatterjee S (1999) *Protoavis* and the early evolution of birds. Palaeontographica A 254:1–100

Chatterjee S, Templin RJ (2007) Biplane wing planform and flight performance of the feathered dinosaur *Microraptor gui*. Proc Natl Acad Sci 104:1576–1580

Chiappe L, Lacasa-Ruiz J (2002) *Noguerornis gonzalezi* (Aves: Ornithothoraces) from the Early Cretaceous of Spain. In: Chiappe LM, Witmer LM (eds) Mesozoic birds: above the heads of dinosaurs. Berkeley University Press, Berkeley, USA, pp 230–239

Christensen P, Bonde N (2004) Body plumage in *Archaeopteryx*: a review, and new evidence from the Berlin specimen. C R Palevol 3:99–118

Currie PJ, Zhiming D (2001) New information on Cretaceous troodontids from the People's Republic of China. Can J Earth Sci 38:1753–1766

Feduccia A (1996) The origin and evolution of birds. Yale University Press, New Haven

Forster C, Sampson S, Chiappe LM, Krause D (1998) The theropod ancestry of birds: new evidence from the Late Cretaceous of Madagascar. Science 279:1915–1919

Foth C (2011) On the identification of feather structures in stem-line representatives of birds: evidence from fossils and actuopalaeontology. Palaontologische Zeitschrift. doi:10.1007/s12542-011-0111-3

Hone DWE, Tischlinger H, Xu X, Zhang F (2010) The extent of the preserved feathers on the four-winged dinosaur *Microraptor gui* under ultraviolet light. PlosOne 5:e9223. doi:10.1371/journal.pone.0009223

Hu D, Hou L, Zhang L, Xu X (2009) A pre-*Archaeopteryx* troodontid theropod from China with long feathers on the metatarsus. Nature 461:640–643

Jasinoski SC, Russell AP, Currie PJ (2006) An integrative phylogenetic and extrapolatory approach to the reconstruction of dromaeosaur (Theropoda: Eumaniraptora) shoulder musculature. Zool J Linn Soc 146:301–344

Kumar S, Hedges B (1998) A molecular timescale for vertebrate evolution. Nature 392:917–920

Makovicky PJ, Apesteguía S, Agnolín FL (2005) The earliest dromaeosaurid theropod from South America. Nature 437:1007–1011

Martin LD, Lim JD (2005) Soft body impression of the hand in *Archaeopteryx*. Curr Sci 89:1089–1090

Novas FE (2009) The Age of Dinosaurs in South America. Indiana University Press, Indiana, pp 452

Novas FE, Puerta P (1997) New evidence concerning avian origins from the Late Cretaceous of NW Patagonia. Nature 387:390–392

Ortega F, Escaso F, Sanz JL (2010) A bizarre, humped Carcharodontosauria (Theropoda) from the Lower Cretaceous of Spain. Nature 467:203–206

Ostrom JH (1969) Osteology of *Deinonychus antirrhopus*, an unusual theropod from the Lower Cretaceous of Montana. Bull Peabody Mus Nat Hist 30:1–165

Ostrom JH (1976) *Archaeopteryx* and the origin of birds. Biol J Linn Soc 8:91–182

Paul GS (2002) Dinosaurs of the air. The John Hopkins University Press

Sanz JL, Chiappe LM, Pérez-Moreno BP, Buscalioni AD, Moratalla JJ, Ortega F, Poyato-Ariza FJ (1996) An Early Cretaceous bird from Spain and its implications for the evolution of avian flight. Nature 382:442–445

Senter P (2006) Scapular orientation in theropods and basal birds, and the origin of flapping flight. Act Palaeont Pol 51:305–313

Senter P (2007) A new look at the phylogeny of Coelurosauria (Dinosauria: Theropoda). J Syst Palaeont 5:429–463

Senter P, Barsbold R, Britt B, Burnham D (2004) Systematics and evolution of Dromaeosauridae (Dinosauria: Theropoda). Bull Gunma Mus Nat Hist 8:1–20

Sereno PC (2004) Birds as dinosaurs. Acta Zool Sin 50:991–1001

Turner AH, Pol D, Clarke JA, Erickson GM, Norell MA (2007a) A basal dromaeosaurid and size evolution preceding avian flight. Science 317:1378–1381

Turner AH, Makovicky PJ, Norel MA (2007b) Feather quill knobs in the dinosaur *Velociraptor*. Science 21317:1721

Turner AH, Pol D, Norell MA (2011) Anatomy of *Mahakala omnogovae* (Theropoda: Dromaeosauridae), Tögrögiin Shiree, Mongolia. Am Museum Novitates 3722:1–66

Vasquez RJ (1992) Functional osteology of the avian wrist and the evolution of flapping flight. J Morph 211:259–268

Vasquez RJ (1994) The automating skeletal and muscular mechanisms of the avian wing (Aves). Zoomorph 114:59–71

Witmer LM (2009) Palaeontology: feathered dinosaurs in a tangle. Nature 461:601–602

Xu X (2002) Deinonychosaurian fossils from the Jehol Group of western Liaoning and the coelurosaurian evolution. Dissertation for the doctoral degree. Chinese Academy of Sciences, Beijing

Xu X, Guo Y (2009) The origin and early evolution of feathers: insights from recent paleontological and neontological data. Vert Palas 47:311–329

Xu X, Zhang F (2005) A new maniraptoran dinosaur from China with long feathers on the metatarsus. Naturwissenschaften 92:173–177

Xu X, Wang X-L, Wu X-C (1999) A dromaeosaurid dinosaur with a filamentous integument from the Yixian Formation of China. Nature 401:262–266

Xu X, Zhou Z, Wang X (2000) The smallest known non-avian theropod dinosaur. Nature 408:705–708

Xu X, Zhou Z, Wang X, Huang X, Zhang F, Du X (2003) Four winged dinosaurs from China. Nature 421:335–340

Xu X, Zhao Q, Norell MA, Sullivan C, Hone D, Erickson PG, Wang X, Han F, Guo Y (2008) A new feathered dinosaur fossil that fills a morphological gap in avian origin. Chinese Sci Bull 54:430–435

Xu X, Zheng X, You H (2010) Exceptional dinosaur fossils show ontogenetic development of early feathers. Nature 464:1338–1341

Xu X, You H, Du K, Han F (2011) An *Archaeopteryx*-like theropod from China and the origin of Avialae. Nature 475:465–470

Zhang F (2004) The origin and early evolution of birds: discoveries, disputes, and perspectives from fossil evidence. Naturwissenschaften 91:455–471

Zhang F, Zhou Z (2004) Leg feathers in an early cretaceous bird. Nature 431:925

Zhang F, Zhou Z, Xu X, Wang X, Sullivan C (2008) A bizarre Jurassic maniraptoran from China with elongate ribbon-like feathers. Nature 455:1105–1108

Zheng X, Xu X, You H, Zhao Q, Dong Z (2009) A short-armed dromaeosaurid from the Jehol Group of China with implications for early dromaeosaurid evolution. Proc Royal Soc London B 277:211–217

Zhou Z, Zhang F (2006) A beaked basal ornithurine bird (Aves, Ornithurae) from the Lower Cretaceous of China. Zool Script 35:363–373

Appendix 1
Character List

1. Vaned feathers on forelimb symmetric (0) or asymmetric (1).
2. Orbit round in lateral or dorsolateral view (0) or dorsoventrally elongate (1).
3. Anterior process of postorbital projects into orbit (0) or does not project into orbit (1).
4. Postorbital in lateral view with subhorizontal anterior (frontal) process (0) or frontal process diagonal (anterior tip of process higher than base of process) (1).
5. Postorbital bar parallels quadrate, lower temporal fenestra rectangular in shape (0) or jugal and postorbital approach or contact quadratojugal to constrict lower temporal fenestra (1).
6. Otosphenoidal crest vertical on basisphenoid and prootic, and does not border an enlarged pneumatic recess (0) or well developed, crescent shaped, thin crest forms anterior edge of enlarged pneumatic recess (1).
7. Crista interfenestralis confluent with lateral surface of prootic and opisthotic (0) or distinctly depressed within middle ear opening (1).
8. Subotic recess (pneumatic fossa ventral to fenestra ovalis) absent (0) or present (1).
9. Basisphenoid recess present between basisphenoid and basioccipital (0) or entirely within basisphenoid (1) or absent (2).
10. Posterior opening of basisphenoid recess single (0) or divided into two small, circular foramina by a thin bar of bone (1).
11. Base of cultriform process not highly pneumatized (0) or base of cultriform process (parasphenoid rostrum) expanded and pneumatic (parasphenoid bulla) (1).
12. Basipterygoid processes ventral or anteroventrally projecting (0) or lateroventrally projecting (1).
13. Basipterygoid processes well developed, extending as a distinct process from the base of the basisphenoid (0) or processes abbreviated or absent (1).
14. Basipterygoid processes solid (0) or processes hollow (1).
15. Basipterygoid recesses on dorsolateral surfaces of basipterygoid processes absent (0) or present (1).

F. L. Agnolín and F. E. Novas, *Avian Ancestors*, SpringerBriefs in Earth System Sciences, DOI: 10.1007/978-94-007-5637-3, © The Author(s) 2013

16. Depression for pneumatic recess on prootic absent (0) or present as dorsally open fossa on prootic/opisthotic (1) or present as deep, posterolaterally directed concavity (2).
17. Accessory tympanic recess dorsal to crista interfenestralis absent (0) small pocket present (1) or extensive with indirect pneumatization (2).
18. Caudal (posterior) tympanic recess absent (0) present as opening on anterior surface of paroccipital process (1) or extends into opisthotic posterodorsal to fenestra ovalis, confluent with this fenestra (2).
19. Exits of C. N. X–XII flush with surface of exoccipital (0) or cranial nerve exits located together in a bowl-like basisphenoid depression (1).
20. Maxillary process of premaxilla contacts nasal to form posterior border of nares (0) or maxillary process reduced so that maxilla participates broadly in external naris (1) or maxillary process of premaxilla extends posteriorly to separate maxilla from nasal posterior to nares (2).
21. Internarial bar rounded (0) or flat (1).
22. Crenulate margin on buccal edge of premaxilla absent (0) or present (1).
23. Caudal margin of naris farther rostral than (0), or nearly reaching or overlapping (1), the rostral border of the antorbital fossa.
24. Premaxillary symphysis acute, V-shaped (0) or rounded, U-shaped (1).
25. Secondary palate short (0) or long, with extensive palatal shelves on maxilla (1).
26. Palatal shelf of maxilla flat (0) or with midline ventral 'tooth-like' projection (1).
27. Pronounced, round accessory antorbital fenestra absent (0) or present (1).
28. Accessory antorbital fossa situated at rostral border of antorbital fossa (0) or situated posterior to rostral border of fossa (1).
29. Tertiary antorbital fenestra (fenestra promaxillaris) absent (0) or present (1).
30. Antorbital fossa without distinct rim ventrally and anteriorly (0) or with distinct rim composed of a thin wall of bone (1).
31. Narial region apneumatic or poorly pneumatized (0) or with extensive pneumatic fossae, especially along posterodorsal rim of fossa (1).
32. Jugal and postorbital contribute equally to postorbital bar (0) or ascending process of jugal reduced and descending process of postorbital ventrally elongate (1).
33. Jugal quadratojugal process tall beneath lower temporal fenestra, twice or more as tall dorsoventrally as it is wide transversely (0) or rod-like (1) or concealed by quadratojugal (2).
34. Jugal pneumatic recess in posteroventral corner of antorbital fossa present (0) or absent (1).
35. Medial jugal foramen present on medial surface ventral to postorbital bar (0) or absent (1).
36. Quadratojugal without horizontal process posterior to ascending process (reversed "L" shape) (0) or with process (i.e., inverted 'T' or 'Y' shape) (1).
37. Jugal and quadratojugal separate (0) or quadratojugal and jugal fused and not distinguishable from one another (1).
38. Supraorbital crests on lacrimal in adult individuals absent (0) or dorsal crest above orbit (1) or lateral expansion anterior and dorsal to orbit (2).

39. Enlarged foramen or foramina opening laterally at the angle of the lacrimal, absent (0) or present (1).
40. Lacrimal posterodorsal process absent (inverted 'L' shaped) (0) or lacrimal 'T' shaped in lateral view (1) or anterodorsal process much longer than posterior process (2) or posterodorsal process subvertical (3).
41. Prefrontal large, dorsal exposure similar to that of lacrimal (0) or greatly reduced in exposure (1) or without exposure (2).
42. Frontals narrow anteriorly as a wedge between nasals (0) or end abruptly anteriorly, suture with nasal transversely orientated (1) or suture with nasals W-shaped (2).
43. Anterior emargination of supratemporal fossa on frontal straight or slightly curved (0) or strongly sinusoidal and reaching onto postorbital process (1).
44. Frontal postorbital process (dorsal view): smooth transition from orbital margin (0) or sharply demarcated from orbital margin (1).
45. Frontal edge smooth in region of lacrimal suture (0) or edge notched (1).
46. Dorsal surface of parietals flat, lateral ridge borders supratemporal fenestra (0) or parietals dorsally convex with very low sagittal crest along midline (1) or dorsally convex with well developed sagittal crest (2).
47. Parietals separate (0) or fused (1).
48. Descending process of squamosal parallels quadrate shaft (0) or nearly perpendicular to quadrate shaft (1).
49. Descending process of squamosal contacts quadratojugal (0) or does not contact quadratojugal (1).
50. Posterolateral shelf on squamosal overhanging quadrate head absent (0) or present (1).
51. Dorsal process of quadrate single headed (0) or with two distinct heads, a lateral one contacting the squamosal and a medial head contacting the braincase (1).
52. Quadrate vertical (0) or strongly inclined anteroventrally so that distal end lies far forward of proximal end (1).
53. Quadrate solid (0) or hollow, with depression on posterior surface (1).
54. Lateral border of quadrate shaft straight (0) or with lateral tab that touches squamosal and quadratojugal above an enlarged quadrate foramen (1).
55. Foramen magnum subcircular, slightly wider than tall (0) or oval, taller than wide (1).
56. Occipital condyle without constricted neck (0) or subspherical with constricted neck (1).
57. Paroccipital process elongate and slender, with dorsal and ventral edges nearly parallel (0) or process short, deep with convex distal end (1).
58. Paroccipital process straight, projects laterally or posterolaterally (0) or distal end curves ventrally, pendant (1).
59. Paroccipital process with straight dorsal edge (0) or with dorsal edge twisted rostrolaterally at distal end (1).
60. Ectopterygoid with constricted opening into fossa (0) or with open ventral fossa in the main body of the element (1).
61. Dorsal recess on ectopterygoid absent (0) or present (1).

62. Flange of pterygoid well developed (0) or reduced in size or absent (1).
63. Palatine and ectopterygoid separated by pterygoid (0) or contact (1).
64. Palatine tetraradiate, with jugal process (0) or palatine triradiate, jugal process absent (1).
65. Suborbital fenestra similar in length to orbit (0) or about half or less than half orbital length (1) or absent (2).
66. Symphyseal region of dentary broad and straight, paralleling lateral margin (0) or medially recurved slightly (1) or strongly recurved medially (2).
67. Dentary symphyseal region in line with main part of buccal edge (0) or abruptly downturned at rostral end (1) or dentary ramus gradually, weakly downturned through its length (2).
68. Mandible without coronoid prominence (0) or with coronoid prominence (1).
69. Posterior end of dentary without posterodorsal process dorsal to mandibular fenestra (0) or with dorsal process above anterior end of mandibular fenestra (1) or with elongate, strongly arched dorsal process extending over most of fenestra (2).
70. Labial face of dentary flat (0) or with lateral ridge and inset tooth row (1).
71. Dentary subtriangular in lateral view (0) or with subparallel dorsal and ventral edges (1).
72. Nutrient foramina on external surface of dentary superficial (0) or lie within deep groove (1).
73. External mandibular fenestra oval (0) or subdivided by a spinous rostral process of the surangular (1).
74. Internal mandibular fenestra small and slit-like (0) or large and rounded (1).
75. Foramen in lateral surface of surangular rostral to mandibular articulation, absent (0) or present (1).
76. Splenial not widely exposed on lateral surface of mandible (0) or exposed as a broad triangle between dentary and angular on lateral surface of mandible (1).
77. Coronoid ossification large (0) or only a thin splint (1) or absent (2).
78. Articular without elongate, slender medial, posteromedial, or mediodorsal process from retroarticular process (0) or with process (1).
79. Retroarticular process short, stout (0) or elongate and slender (1).
80. Mandibular articulation surface as long as distal end of quadrate (0) or twice or more as long as quadrate surface, allowing anteroposterior movement of mandible (1).
81. Premaxilla toothed (0) or edentulous (1).
82. Second premaxillary tooth approximately equivalent in size to other premaxillary teeth (0) or second tooth markedly larger than third and fourth premaxillary teeth (1) or first premaxillary tooth huge, other premaxillary teeth tiny (2) or first premaxillary tooth larger than the others but all premaxillary teeth tiny (3).
83. Maxilla toothed (0) or edentulous (1).
84. Maxillary and dentary teeth serrated (0) or some without serrations anteriorly (except at base in *S. mongoliensis*) (1) or all without serrations (2).

85. Dentary and maxillary teeth large, less than 25 in dentary (0) or large number of small teeth (25 or more in dentary) (1) or small number of dentary teeth ($\leq$11) (2) or dentary without teeth (3).
86. Serration denticles large (0) or small (1).
87. Serrations simple, denticles convex (0) or distal and often mesial edges of teeth with large, hooked denticles that point toward the tip of the crown (1).
88. Teeth constricted between root and crown (0) or root and crown confluent (1).
89. Dentary teeth evenly spaced (0) or anterior dentary teeth smaller, more numerous, and more closely appressed than those in middle of tooth row (1).
90. Dentaries lack distinct interdental plates (0) or with interdental plates medially between teeth (1).
91. In cross section, premaxillary tooth crowns sub-oval to sub-circular (0) or asymmetrical (D-shaped in cross section) with flat lingual surface (1) or first premaxillary tooth with flat lingual surface, other premaxillary teeth without flat lingual surfaces (2).
92. Number of cervical vertebrae: 10 (0) or 12 or more (1).
93. Axial epipophyses absent or poorly developed, not extending past posterior rim of postzygapophyses (0) or large and posteriorly directed, extend beyond postzygapophyses (1).
94. Axial neural spine flared transversely (0) or compressed mediolaterally (1).
95. Epipophyses of cervical vertebrae placed distally on postzygapophyses, above postzygapophyseal facets (0) or placed proximally, proximal to postzygapophyseal facets (1).
96. Anterior cervical centra level with or shorter than posterior extent of neural arch (0) or centra extending beyond posterior limit of neural arch (1).
97. Carotid process on posterior cervical vertebrae absent (0) or present (1).
98. Anterior cervical centra subcircular or square in anterior view (0) or distinctly wider than high, kidney shaped (1).
99. Cervical neural spines anteroposteriorly long and dorsoventrally tall (0) or anteroposteriorly short, dorsoventrally low and centred on neural arch, giving arch an 'X' shape in dorsal view (1) or anteroposteriorly short and dorsoventrally tall (2) or anteroposteriorly long and dorsoventrally short (3).
100. Cervical centra with one pair of pneumatic openings (0) or with two pairs of pneumatic openings (1).
101. Cervical and anterior trunk vertebrae amphiplatyan (0) or opisthocoelous (1).
102. Anterior trunk vertebrae without prominent hypapophyses (0) or with large hypapophyses (1).
103. Parapophyses of posterior trunk vertebrae flush with neural arch (0) or distinctly projected on pedicels (1).
104. Hyposphene-hypantrum articulations in trunk vertebrae absent (0) or present (1).
105. Zygapophyses of trunk vertebrae abutting one another above neural canal, opposite hyposphenes meet to form lamina (0), or zygapohyses placed lateral to neural canal and separated by groove for interspinuous ligaments, hyposphenes separated (1).

106. Middle and posterior dorsal vertebrae not pneumatic (0) or pneumatic (1).
107. Transverse processes of anterior dorsal vertebrae long and thin (0) or short, wide, and only slightly inclined (1).
108. Neural spines of dorsal vertebrae not expanded distally (0) or expanded to form 'spine table' (1).
109. Scars for interspinous ligaments terminate at apex of neural spine in dorsal vertebrae (0) or terminate below apex of neural spine (1).
110. Number of sacral vertebrae: 5 (0) or 6 (1) or 7 or more (2).
111. Sacral vertebrae with unfused zygapophyses (0) or with fused zygapophyses forming a sinuous ridge in dorsal view (1).
112. Ventral surface of posterior sacral centra gently rounded, convex (0) or ventrally flattened, sometimes with shallow sulcus (1) or centrum strongly constricted transversely, ventral surface keeled (2).
113. Pleurocoels absent on sacral vertebrae (0) or present on anterior sacrals only (1) or present on all sacrals (2).
114. Last sacral centrum with flat posterior articulation surface (0) or convex articulation surface (1).
115. Caudal vertebrae with distinct transition point (0) or without transition point (1).
116. Transition point in caudal series begins distal to the 10th caudal (0) or between 7th and 10th caudal vertebra (1) or proximal to the 7th caudal vertebra (2).
117. Anterior caudal centra tall, oval in cross section (0) or with box-like centra in caudals I-V (1) or anterior caudal centra laterally compressed with ventral keel (2).
118. Neural spines of caudal vertebrae simple, undivided (0) or separated into anterior and posterior alae throughout much of caudal sequence (1).
119. Neural spines on distal caudals form a low ridge (0) or spine absent (1) or midline sulcus in center of neural arch (2).
120. Prezygapophyses of distal caudal vertebrae between 1/3 and whole centrum length (0) or with extremely long extensions of the prezygapophyses (up to 10 vertebral segments long in some taxa) (1) or strongly reduced as in *Archaeopteryx lithographica* (2).
121. More than 30 caudal vertebrae (0) or 21–30 caudal vertebrae (1) or <10 caudal vertebrae, followed by pygostyle (2) or 11–20 vertebrae (3).
122. Proximal end of chevrons of proximal caudals short anteroposteriorly, shaft proximodistally elongate (0) or proximal end elongate anteroposteriorly, flattened and plate-like (1).
123. Distal caudal chevrons are simple (0) or anteriorly bifurcate (1) or bifurcate at both ends (2).
124. Shaft of cervical ribs slender and longer than vertebra to which they articulate (0) or broad and shorter than vertebra (1).
125. Ossified uncinate processes absent (0) or present (1).
126. Ossified ventral rib segments absent (0) or present (1).
127. Lateral gastral segment shorter than medial one in each arch (0) or distal segment longer than proximal segment (1).

128. Ossified sternal plates separate in adults (0) or fused (1).

129. Sternum without distinct lateral xiphoid process posterior to costal margin (0) or with lateral xiphoid process (1).

130. Anterior edge of sternum grooved for reception of coracoids (0) or sternum without grooves (1).

131. Articular facet of coracoid on sternum (conditions may be determined by the articular facet on coracoid in taxa without ossified sternum): anterolateral or more lateral than anterior (0); almost anterior (1).

132. Hypocleidium on furcula absent (0) or present (1).

133. Acromion margin of scapula continuous with blade (0) or anterior edge laterally everted (1).

134. Anterior surface of coracoid ventral to glenoid fossa unexpanded (0) or anterior edge of coracoid expanded, forms triangular subglenoid fossa bounded laterally by coracoid tuber (1).

135. Scapula and coracoid separate (0) or fused into scapulacoracoid (1).

136. Coracoid in lateral view subcircular, with shallow ventral blade (0) or subquadrangular with extensive ventral blade (1) or shallow ventral blade with elongate posteroventral process (2) or subtriangular (proximal end constricted, distal end wide) (3).

137. Scapula and coracoid form a continuous arc in posterior and anterior views (0) or coracoid inflected medially, scapulocoracoid 'L' shaped in lateral view (1).

138. Glenoid fossa without (0) or with extension of glenoid floor onto external surface of scapula (the surface opposite the costal surface) (1).

139. Scapula longer than humerus (0) or humerus longer than scapula (1).

140. Deltopectoral crest large and distinct, proximal end of humerus quadrangular in anterior view (0) or deltopectoral crest less pronounced, forming an arc rather than being quadrangular (1) or deltopectoral crest very weakly developed, proximal end of humerus with rounded edges (2) or deltopectoral crest extremely long (3) or proximal end of humerus extremely broad, triangular in anterior view (4).

141. Anterior surface of deltopectoral crest smooth (0) or with distinct groove or ridge near lateral edge along distal end of crest (1).

142. Olecranon process weakly developed (0) or distinct and large but not hypertrophied (1) or hypertrophied (2).

143. Articulation between ulna and radius flat (0) or peg and socket (1).

144. Proximal surface of ulna a single continuous articular facet (0) or divided into two distinct fossae separated by a median ridge (1).

145. Lateral proximal carpal (ulnare?) quadrangular (0) or triangular in proximal view (1).

146. Two distal carpals in contact with metacarpals, one covering the base of metacarpal I (and perhaps contacting metacarpal II) the other covering the base of metacarpal II (distal carpals 1 and 2 unfused) (0) or a single distal carpal capping metacarpals I and II (distal carpals 1 and 2 fused) (1).

147. Distal carpals not fused to metacarpals (0) or fused to metacarpals, forming carpometacarpus (1).

148. Distal carpals 1 + 2 well developed, covering all of proximal ends of meta-carpals I and II (0) or small, cover about half of base of metacarpals I and II (1) or cover bases of all metacarpals (2).
149. Metacarpal I half or less than half the length of metacarpal II, and longer proximodistally than wide transversely (0) or subequal in length to metacar-pal II (1) or very short and wider transversely than long proximodistally (2).
150. Third manual digit present, phalanges present (0) or reduced to no more than metacarpal splint (1).
151. Flexor tubercles of manual unguals proximal (0) or displaced distally from articular end (1) or proximodistally elongated with proximal end close to articular facet (2).
152. Unguals on all digits generally similar in size (0) or digit I bearing large ungual and unguals of other digits distinctly smaller (1).
153. Proximodorsal 'lip' on first manual ungual—a transverse ridge immediately dorsal to the articulating surface—absent (0) or present (1).
154. Ventral edge of anterior ala of ilium straight or gently curved (0) or ventral edge hooked anteriorly (1) or very strongly hooked (2).
155. Preacetabular part of ilium roughly as long as postacetabular part of ilium (0) or preacetabular portion of ilium markedly longer (more than 2/3 of total ilium length) than postacetabular part (1).
156. Anterior end of ilium gently rounded or straight (0) or anterior end strongly curved (1) or pointed at anterodorsal corner (2).
157. Supraacetabular crest on ilium as a separate process from antitrochanter, forms "hood" over femoral head present (0) reduced, not forming hood (1) or absent (2).
158. Postacetabular ala of ilium in lateral view squared (0) or acuminate (1).
159. Postacetabular blades of ilia in dorsal view parallel (0) or diverge posteriorly (1).
160. Tuber along dorsal edge of ilium, dorsal or slightly posterior to acetabulum absent (0) or present (1).
161. Brevis fossa shelf-like (0) or deeply concave with lateral overhang (1).
162. Antitrochanter posterior to acetabulum absent or poorly developed (0) or prominent (1).
163. Ridge bordering cuppedicus fossa extends far posteriorly and is confluent or almost confluent with acetabular rim (0) or ridge terminates rostral to acetab-ulum or curves ventrally onto anterior end of pubic peduncle (1).
164. Cuppedicus fossa deep, ventrally concave (0) or fossa shallow or flat, with no lateral overhang (1) or absent (2).
165. Posterior edge of ischium without (0) or with prominent proximodorsal prong (1).
166. Shaft of ischium straight in lateral view (0) or ventrodistal end curved anteri-orly (1) or curved dorsally (2).
167. Obturator process of ischium absent (0) or proximal in position (1) or dis-tally displaced (2).
168. Obturator process does not contact pubis (0) or contacts pubis (1).

169. Length of pubic boot ≤30 % length of pubis (0) or ≥40 % (1).
170. Semicircular scar on posterior part of the proximal end of the ischium, absent (0) or present (1).
171. Ischium more than 70 % (0) or 70 % or less of pubis length (1).
172. Distal ends of ischia form symphysis (0) or approach one another but do not form symphysis (1) or widely separated (2).
173. Ischial boot (expanded distal end) present (0) or absent (1).
174. Tubercle on anterior edge of ischium absent (0) or present (1).
175. Pubis propubic (0) or pubis vertical (1) or pubis moderately posteriorly oriented (2) or pubis fully posteriorly oriented (opisthopubic) (3).
176. Pubic boot projects anteriorly and posteriorly (0) or with little or no anterior process (1) or no anteroposterior projections (2).
177. Shelf on pubic shaft proximal to symphysis (pubic apron) extends medially from middle of cylindrical pubic shaft (0) or shelf extends medially from anterior edge of anteroposteriorly flattened shaft (1).
178. Pubic shaft straight (0) or distal end curves anteriorly, anterior surface of shaft concave in lateral view (1) or anterior surface of shaft convex in lateral view (2).
179. Pubic apron about half of pubic shaft length (0) or less than 1/3 of shaft length (1).
180. Femoral head without fovea capitalis (for attachment of capital ligament) (0) or circular fovea present in center of medial surface of head (1).
181. Lesser and greater trochanters unfused (0) or fused (1).
182. Lesser trochanter of femur alariform (0) or cylindrical in cross section (1).
183. Posterior trochanter absent or represented only by rugose area (0) or posterior trochanter distinctly raised from shaft, mound-like (1).
184. Fourth trochanter on femur present (0) or absent (1).
185. Accessory trochanteric crest distal to lesser trochanter absent (0) or present (1).
186. Anterior surface of femur proximal to medial distal condyle without longitudinal crest (0) or crest present extending proximally from medial condyle on anterior surface of shaft (1).
187. Popliteal fossa on distal end of femur open distally (0) or closed off distally by contact between distal condyles (1).
188. Fibula reaches proximal tarsals (0) or short, tapering distally, and not in contact with proximal tarsals (1).
189. Medial surface of proximal end of fibula concave along long axis (0) or flat (1).
190. Deep oval fossa on medial surface of fibula near proximal end absent (0) or present (1).
191. Distal end of tibia and astragalus without distinct condyles (0) or with distinct condyles separated by prominent tendinal groove on anterior surface (1).
192. Medial cnemial crest absent (0) or present on proximal end of tibia (1).
193. Ascending process of the astragalus tall and broad, covering most of anterior surface of distal end of tibia (0) or process short and slender, covering only lateral half of anterior surface of tibia (1) or ascending process tall with medial notch that restricts it to lateral side of anterior face of distal tibia (2).

194. Ascending process of astragalus confluent with condylar portion (0) or separated by transverse groove or fossa across base (1).
195. Astragalus and calcaneum separate from tibia (0) or fused to each other and to the tibia in late ontogeny (1).
196. Distal tarsals separate, not fused to metatarsals (0) or form metatarsal cap with intercondylar prominence that fuses to metatarsal early in postnatal ontogeny (1).
197. Metatarsals not co-ossified (0) or co-ossification of metatarsals begins proximally (1) or distally (2).
198. Distal end of metatarsal II smooth, not ginglymoid (0) or with developed ginglymus (1).
199. Distal end of metatarsal III smooth, not ginglymoid (0) or with developed ginglymus delimited proximally by a ridge (1) or with a poorly developed ginglymus with proximal ridge absent (2).
200. In anterior view, metatarsal III not pinched (0) or pinched proximally (1) or pinched both proximally and through midshaft (2).
201. Ungual of pedal digit II similar in size to that of III (0) or pedal ungual II about 50% larger than pedal ungual III (1).
202. Metatarsal I articulates at middle of metatarsal II (0) or metatarsal I attaches to distal quarter of metatarsal II (1) or metatarsal I articulates with metatarsal II near its proximal end (2) or metatarsal I absent (3).
203. Metatarsal I attenuates proximally (0) or proximal end of metatarsal I similar to that of metatarsals II–IV (1).
204. Shaft of MT IV round or thicker dorsoventrally than wide in cross section (0) or shaft of MT IV mediolaterally widened and flat in cross section (1).
205. Foot symmetrical (0) or asymmetrical with slender MTII and very robust MT IV (1).
206. Neural spines on posterior dorsal vertebrae in lateral view rectangular or square (0) or anteroposteriorly expanded distally, fanshaped (1).
207. Shaft diameter of phalanx I–1 less (0) or greater (1) than shaft diameter of radius.
208. Angular exposed almost to end of mandible in lateral view, reaches or almost reaches articular (0) or excluded from posterior end angular suture turns ventrally and meets ventral border of mandible rostral to glenoid (1).
209. Laterally inclined flange along dorsal edge of surangular for articulation with lateral process of lateral quadrate condyle absent (0) or present (1).
210. Distal articular ends of metacarpals I + II ginglymoid (0) or rounded, smooth (1).
211. Radius and ulna well separated (0) or with distinct adherence or syndesmosis distally (1).
212. Kink and downward deflection in dentary buccal margin at rostral end of dentary: absent (0) or present (1).
213. Quadrate head covered by squamosal in lateral view (0) or quadrate cotyle of squamosal open laterally exposing quadrate head (1).
214. Brevis fossa poorly developed adjacent to ischial peduncle and without lateral overhang, medial edge of brevis fossa visible in lateral view (0), or fossa well developed along full length of postacetabular blade, lateral overhang

extends along full length of fossa, medial edge completely covered in lateral view (1).

215. Vertical ridge on lesser trochanter present (0) or absent (1).

216. Supratemporal fenestra bounded laterally and posteriorly by the squamosal (0) or supratemporal fenestra extended as a fossa on to the dorsal surface of the squamosal (1).

217. Dentary fully toothed (0) or only with teeth rostrally (1) or edentulous (2).

218. Posterior edge of coracoid not or shallowly indented below glenoid (0), or posterior edge of coracoid deely notched just ventral to glenoid, glenoid lip everted (1).

219. Retroarticular process points caudally (0) or curves gently dorsocaudally (1).

220. Flange on supraglenoid buttress on scapula absent (0) or present (1).

221. Depression (possibly pneumatic) on ventral surface of postorbital process of laterosphenoid absent (0) or present (1).

222. Basal tubera set far apart, level with or beyond lateral edge of occipital condyle and/or foramen magnum (may connected by a web of bone or separated by a large notch) (0) or tubera small, directly below condyle and foramen magnum, and separated by a narrow notch (1).

223. Basioccipital without pneumatization on occipital surface (0) or with subcondylar recess (1).

224. Ventral surface of dentary straight or nearly straight (0) or descends strongly posteriorly (1).

225. Distal humerus with small or no medial epicondyle (0) or with large medial epicondyle, medial condyle centered on distal end (1).

226. Distal humeral condyles on distal end (0) or on anterior surface (1).

227. Ilium and ischium articulation flat or slightly concavo-convex (0) or ilium with process projecting into socket in ischium (1).

228. Roots of dentary and maxillary teeth mediolaterally compressed (0) or circular in cross-section (1).

229. Preacetabular portion of ilium parasagital (0) moderately laterally flaring (1) strongly laterally flaring (2).

230. Maxillary and dentary teeth labiolingually flattened and recurved, with crowns in middle of tooth row more than twice as high as the basal mesiolateral width (0) or lanceolate and subsymmetrical (1) or conical (2) or labiolingually flattened and recurved, with crowns in middle of tooth row less than twice as high as the basal mesiolateral width (fore-aft basal length) (3).

231. Dentary teeth do not (0) or do increase in size anteriorly, becoming more conical in shape (1).

232. Length of skull more than 90 % femoral length (0) or less than 80 % (1).

233. Height of skull (minus mandible) at middle of naris more than half the height of skull at middle of orbit (0) or less than half (1).

234. Dorsal margin of naris below level of dorsal margin of orbit (0) or above (1).

235. Snout does not (0) or does taper to an anterior point (1).

236. Area of antorbital fenestra greater than that of orbit (0) or less than that of orbit (1).

237. Body of premaxilla dorsoventrally deep (0) or dorsoventrally shallow (1).
238. Antorbital fossa anteriorly bounded by maxilla (0) or by premaxilla (1).
239. Maxillary antorbital fossa: small, from 10 % to less than 40 % of the rostrocaudal length of the antorbital cavity (0), large, greater than 40 % of the rostrocaudal length of the antorbital cavity (1).
240. Maxillary fenestra subhorizontally positioned (0), or dorsally displaced (1).
241. Nasal fusion: absent, nasals separate (0) or present, nasals fused together (1).
242. Nasal surface: smooth (0) or rugose (1).
243. Suborbital process of jugal short and dorsoventrally stout (0) or elongate and dorsoventrally narrow (1).
244. Nasals at least as long as frontals (0) or shorter than frontals (1).
245. Anterior upturning of nasals absent (0) or present (1).
246. Jugo-maxillary bar at ventral end of antorbital fenestra dorsoventrally deep (0) or dorsoventrally narrow (1).
247. Anteroventral corner of premaxilla does not (0) or does form an acute, ventrally orientated point in lateral view (1).
248. Length of preorbital region of cranium > height at anterior edge of preorbital bar (exclusive of midline sagittal ridge, if any) (0) or ≤height at anterior edge of preorbital bar (1).
249. Frontals without supraorbital rim (0) or with supraorbital rim (1).
250. Parietals shorter than frontals (0) or longer (1).
251. Length of ventral border of infratemporal fenestra comparable to that of orbit (0) or much shorter (1).
252. Foramen magnum smaller than or subequal to size of occipital condyle (0) or larger than occipital condyle (1).
253. Dentary not bowed (0) or bowed (concave dorsally) (1).
254. Meckelian groove of dentary deep (0) or shallow (1).
255. Dentary without posteroventral process extending to posterior end of external mandibular fenestra (0) or with such a process (1).
256. Horizontal shelf on the lateral surface of the surangular, rostral and ventral to the mandibular condyle: absent or faint ridge (0), prominent and extending laterally (1).
257. Premaxillary teeth subequal in size to (0) or much smaller than (1) the maxillary teeth.
258. Approximately the same number of denticles per 5 mm on mesial keels of teeth as on distal keels (0) or markedly more denticles per 5 mm on mesial keels (1).
259. Maxillary teeth subperpendicular to ventral margin of maxilla (0) or strongly inclined (1).
260. Dentary tooth implantation: in sockets (0), in paradental groove (1).
261. Dentary dentition continues cranially to tip of dentary (0) or terminates before reaching dentary tip (1).
262. Length of mid-cervical centra approximately the same as dorsal centra (0) or markedly longer than dorsal centra (1).
263. Cervical prezygapophyses unflexed (0) or flexed (1).

264. Dorsal centra $\geq 1.2 \times$ taller than long (0) or height $\leq$ length (1).
265. Posterior dorsal neural spines $\geq 1.5 \times$ taller than long (0) or height $<1.5 \times$ length (1).
266. Postzygapophyses of middle and posterior dorsal vertebrae do not extend posterior to centrum (0) or do (1).
267. Anteriormost haemal arches $\geq 1.5 \times$ longer than associated centra (0) or $<1.5 \times$ as long as centra (1).
268. Angle between furcular arms $>80°$ (0) or $<60°$ (1).
269. Acromion process contacts coracoid (0), or reduced and does not contact coracoid (1).
270. Acromion process does not match any of the following descriptions: (0) rectangular with its dorsal edge forming a 90° angle with the dorsal edge of the scapular blade (1) or a quarter-circle in shape (2) or triangular, with apex pointing away from and subparallel to scapular blade (3).
271. Scapulocoracoid dorsal margin: pronounced notch between the acromion process and the coracoid (0) or margin smooth (1).
272. Wide distal expansion of scapula absent (0) or present (1).
273. Acrocoracoid process absent (0) or present (1).
274. Humeral length is half femoral length or less (0) or shorter than femur but more than half femoral length (1) or longer than femur (2).
275. Length of humeral shaft between deltopectoral crest and distal condyles $<4.5 \times$ shaft diameter (0) or $>4.5 \times$ shaft diameter (1).
276. Ulna not bowed away from humerus (0), or bowed away from humerus (1).
277. Length of radius $<1/3$ femoral length (0) or between 1/3 and 2/3 femoral length (1) or between 2/3 and $1 \times$ femoral length (2) or $>$femoral length (3).
278. Radial diameter $>0.5 \times$ ulnar diameter (0) or $\leq 0.5 \times$ (1).
279. Distal carpals $1 + 2$ flattish (0) or semilunate in shape (1).
280. Length of manual digit II (including metacarpal) $<1.25 \times$ femoral length (0) or $\geq 1.25 \times$ femoral length (1).
281. Distal end of metacarpal I medially (0) or laterally rotated (1).
282. Medial side of metacarpal II: expanded proximally (0), not expanded (1).
283. Metacarpal III $>0.8 \times$ length of metacarpal II (0) or $<0.8 \times$ (1).
284. Manual phalanx I–1 longer than metacarpal II (0) or shorter (1) (P′erez-Moreno et al. 1994).
285. Length of metacarpal II $<$ length of metacarpal I $+$ phalanx I–1 (0) or$\geq$ (1).
286. Metacarpals II and III are not (0) or are appressed for their entire lengths (1).
287. Proximal end of metacarpal III is not (0) or is mainly palmar to that of metacarpal II (1).
288. Length of manual phalanx II–2 $< 1.2 \times$ length of phalanx II–1 (0) or $>1.2 \times$ (1).
289. Medial ligament pits of manual phalanges deep (0) or shallow (1).
290. Posterior flange on manual phalanx II–1 absent (0) or present (1).
291. Combined lengths of manual phalanges II–1 and II–2 $>$ length of metacarpal II $+$ carpus (0) or $\leq$ length of metacarpal II $+$ carpus (1).
292. Length of manual phalanx II–1 $< 2 \times$ length of III–1 (0) or $\geq 2 \times$ length of III–1 (1).

293. Length of manual phalanx II–2 < 2 × length of II–1 (0) or ≥2 × (1).
294. Length of manual phalanx III–1 < 2 × length of phalanx III–2 (0) or >2 × (1).
295. Manual phalanx I–1 straight (0) or bowed (palmar surface concave) (1).
296. With proximal articular surface of ungual orientated vertically, dorsal surface of manual ungual I does not (0) or does arch higher than level of dorsal extremity of proximal articular surface (1).
297. With proximal articular surface of ungual orientated vertically, dorsal surface of manual ungual II does not (0) or does arch higher than level of dorsal extremity of proximal articular surface (1).
298. Manual ungual I strongly curved (0), weakly curved (1), or straight (2).
299. Manual unguals II and III strongly curved (0), weakly curved, (1), or straight (2).
300. Proximodorsal 'lip' on manual unguals II and III absent (0) or present (1).
301. Manual digit III with four phalanges (0) or less than four phalanges (1).
302. Manual phalanx III–3 markedly shorter than combined lengths of phalanges III–1 and III–2 (0), subequal in length to their combined lengths (1), or markedly longer (2).
303. Arching of preacetabular iliac blade above height of postacetabular blade absent or small (0) or extreme (1).
304. Shaft of ischium subequal in thickness to the pubis (0), slenderer than the pubic shaft (1), thicker than the pubic shaft (2).
305. Obturator process does not (0) or does form a strongly acute angle in lateral view (1).
306. Obturator process does not (0) or does reach tip of ischium (1).
307. Ventral notch between the distal portion of the obturator process and the shaft of the ischium: present (0), absent (1).
308. Strong kink of pubis at midshaft absent (0) or present, displacing distal half of pubis caudally (1).
309. In adult, femur longer than tibia (0) or shorter (1).
310. Tip of lesser trochanter below level of femoral head (0) or level with femoral head (1).
311. Proximolateral (fibular) condyle of the tibia, development in proximal view: bulge from the main surface of the tibia (0), conspicuous narrowing between the body of the condyle and the main body of the tibia (1).
312. Metatarsus less than half length of femur (0) or more than half femoral length (1).
313. Metatarsal cross-sectional proportions: subequal or wider mediolaterally than craniocaudally at midshaft (0), deeper craniocaudally than mediolaterally at midshaft (1).
314. Shafts of metatarsals not appressed (0) or appressed (1).
315. Length of metatarsal V ≥0.5 × length of metatarsal IV (0) or <0.5 × (1).
316. Marked decrease in transverse width of metatarsus distally, absent (0) or present (1).
317. Plantar surface of hallux faces posteriorly (0) or hallux reorientated so that plantar surface faces medially or anteriorly (1).

318. Hallucal ungual reduced in size relative to other pedal unguals (0) or not reduced (1).
319. Hallucal ungual weakly curved (0) or strongly curved (1).
320. Length of pedal phalanx II–2 between 0.6 × and 1 × length of phalanx II–1 (0), ≤0.6 ×, or (1) ≥1 × (2).
321. Total length of pedal phalanx II–2 (not counting posteroventral lip, if any) >2 × length of distal condylar eminence (0) or ≤2 × (1).
322. Pedal phalanx II–2 without posteroventral lip or keel (0) with transversely wide posteroventral lip (1) with transversely narrow posteroventral keel (2).
323. Pedal phalanx II–1 without dorsal extension of distal condyles (0) or with extension (1).
324. Pedal unguals III and IV straight or weakly curved (0), or strongly curved (1).
325. With fingers extended, tip of ungual III extends no further distally than flexor tubercle of ungual II (0) or extends further (1).
326. Manual ungual III smaller than ungual II (0) or approximately the same size (1).
327. Diameter of non-ungual phalanges of manual digit III >0.5 × diameter of non-ungual phalanges of digit II (0) or <0.5 × (1).
328. Manual phalanx II–1 shorter than I–1 (0) or longer (1).
329. Ischial shaft rod-like (0) or flat, plate-like (1).
330. Lateral face of ischial shaft flat (or round in rodlike ischia) (0) or laterally concave (1) or with longitudinal ridge dividing lateral surface into anterior and posterior parts (2).
331. Contact between pubic apron contributions of both pubes meet extensively (0) or contact interrupted by a slit (1) or no contact (2).
332. Dorsal margin of postacetabular iliac blade straight or convex (0) or concave (1).
333. Large, longitudinal flange along caudal or lateral face of metatarsal IV absent (0) or present (1).
334. Distally placed dorsal process along caudal edge of ischial shaft absent (0) or present (1).
335. Length of metatarsus <3.5 × transverse midshaft diameter (0) or 3.5–8 × midshaft diameter (1) or >8 × midshaft diameter (2).
336. Lengths of mid-caudal centra subequal to or less than those of proximal caudal centra (0) or ≥twice as long as proximal caudal centra (1).
337. Pubic peduncle of ilium craniocaudally longer (0) or shorter (1) than ischial peduncle of ilium.
338. Phalanges of pedal digit III not blocky (proximal phalanx length ≥2 × diameter) (0) or blocky (proximal phalanx length <2 × diameter) (1).
339. Width of distal humeral expansion<1/3 humeral length (0) or ≥1/3 humeral length (1).
340. Lateral epicondyle of humerus not expanded laterally (0) or expanded laterally (1).
341. Distal end of metatarsal I reduced in size relative to distal ends of other metatarsals (0) or comparable in size to distal ends of other metatarsals (1).

342. Pedal phalanx II–1 longer (0) or shorter (1) than pedal phalanx IV–1.
343. Dentary ramus elongate (0) or shortened, not much longer than tall (1).
344. Metacarpal II ≥1/3 humeral length (0) or <1/3 humeral length (1).
345. With fingers extended, tip of ungual I does not extend past flexor tubercle of ungual II (0) or extends past flexor tubercle of ungual II but does not extend past tip of ungual II (1) or extends past tip of ungual II (2).
346. Premaxillary teeth serrated (0) or unserrated (1).
347. Sublacrimal process of jugal dorsoventrally expanded (taller than suborbital bar of jugal) (0) or not dorsoventrally expanded (1).
348. Flexor tubercles of manual unguals ≥1/3 × height of articular facet (0) or <1/3 (1).
349. Distal chevrons straight or L-shaped in lateral view (0) or upside-down T-shaped (1).
350. Metacarpal III distally not ginglymoid (0) or ginglymoid (1).
351. Breadth of acromion process perpendicular to long axis of scapular blade: deep (0) or shallow (1).
352. Proximal end of metatarsal IV curls around plantar side of proximal end of metatarsal III (0) or does not (1).
353. Midsagittal ridge formed by dorsal displacement of midline of frontals, nasals and premaxillae, absent (0) or present (1).
354. Ectopterygoid lateral to pterygoid (0) or rostral to pterygoid (1).
355. Palatine-pterygoid-ectopterygoid bar does not (0) or does (1) arch below ventral cheek margin.
356. Co-ossification of angular and surangular absent (0) or present (1).
357. Cervical ribs unfused to cervical vertebrae (0) or fused to cervical vertebrae (1).
358. Anteroproximal contact between metatarsals II and IV absent (0) or present (1).
359. Anterior caudal vertebrae without pneumatopores (0) or with pneumatopores (1).
360. External mandibular fenestra not rostrally displaced (sits beneath orbit) (0) or rostrally displaced (sits largely anterior to orbit) (1).
361. Ilium, pubic peduncle: substantially larger than (0) or subequal to (1) ischial peduncle.
362. Ischium, length relative to pubis: shorter (0) or longer (1).
363. Ischium, shape: distally narrower (0) or distally wider (1) (excluding obturator process).
364. Rostral portion of the maxilla elongate: absent (0), present (skull length exceeding femoral length more than 25 %) (1) (Novas et al. 2009).
365. Maxillary fenestra contour: rounded (0), ellipsoidal or slit-like (1) (Novas et al. 2009).
366. Maxilla with postantral wall enlarged: absent (0), present and postantral wall backwardly expanded (1) (Novas et al. 2009).
367. Teeth with longitudinal grooves and ridges: absent (0), present (1). (Gianechini et al. 2009).

368. Posterior margin of ischium: straight or nearly so (0), strongly concave (1) (Zheng et al. 2009).
369. Caudal vertebrae 1–4, transverse process: strap-like, distal portion subequal to or wider anteroposteriorly the base (0) strap-like, distal portion significantly wider anteroposteriorly (1) rode-like, distal end tapered (2).
370. Caudal vertebrae, middle caudal vertebral length: subequal to or slightly longer than (0) or significantly longer than (1) the anteriormost caudal.
371. Scapula, proximal end, medial curvature present, lateral surface of the proximal end significantly medial to that of the scapular bade (0) or absent, about the same level (1).
372. Scapula, articular facet for the coracoid, localization absent (0) present, the dorsal portion of the facet extremely thin transversely, but still on the acromion process (1) present, facet limited ventrally, without development on the acromion process (2).
373. Scapula, acromion process, shape strap-like (0) or ventrally expanded, subtriangular in cross section.
374. Scapula, low crest on the lateral surface of the scapula continuous from the dorsal margin of the acromion process absent (0) present (1).
375. Scapula, acromion process, dorsal extension: present, significant (0) present, minor (1).
376. Scapula, blade robustness strap-like for the distal half, both dorsal and ventral margins sharply ridged (0) relatively robust, only sharply ridged along the dorsal margin close to the distal end (1).
377. Coracoid, distinctly oval-shaped fossa on posterior surface absent (0) present (1).
378. Coracoid, large fenestra absent (0) present (1).
379. Humerus, robustness relative to tibiotarsus significantly more slender than (0) or sub-equal or more robust than (1) tibiotarsus.
380. Humerus, relative length: significantly shorter than (0) or subequal to or longer than (1) femur.
381. Humerus, deltopectoral crest length: long, more than 30 % of the humeral length (0) or short less than 25 % (1).
382. Ulna, robustness relative to tibiotarsus significantly more slender than (0) or more robust than (1) tibiotarsus.
383. Ulna, proximal end, coronoid process prominent (0) or weak (1).
384. Ulna, proximal end, articular surface for ulna condyle flat mediolaterally and longer anteroposteriorly than transversely (0) or a bowl-like fossa, subequal in anteroposterior and mediolateral width (1).
385. Ulna, proximal end, medial process weakly developed (0), or prominent (1).
386. Ulna, a thick ridge along the anterior margin of the proximal third of the shaft absent (0) or present (1).
387. Ulna, distal end, proximal extension of articular facet for the manus along the lateral margin absent or weak (0) or significant (1).
388. Ulna, distal margin of distal end nearly straight (0) or strongly convex (1) in posterior view.

389. Ulna, distal end, anteroposteriorly thickest portion, location: near the medial margin (0) near the mid-length (1).
390. Ulna, radial sulcus weak, shallow depression (0); distinct groove, extending distally to the articular facet (1).
391. Ulna, distal end, anteroposterior flattening absent, transverse width subequal to anteroposterior length (0), present, weak, transverse width to anteroposterior length ratio significantly smaller than 2 (1), present, strong, more than 2 (2).
392. Radius, distal end, lateral flange: present (0), absent (1).
393. Radiale, size small (0) or enlarged (1).
394. Ulnare, shape triangular (0) or V-shaped, with prominent slot (1).
395. 'Semilunate' carpal, position: medially positioned (0); or laterally shifted, centered on metacarpal II (1).
396. Manus, relative length subequal to or less than (0) or significantly longer than (1) femur.
397. Manus, digit III: phalanx III–2 sub-equal to III–1 (0) or significantly shorter than III–1 (1) or significantly longer than III–1 (2).
398. Ilium, anteroventral process, location anteriorly located, close to the anterior extremity of the ilium (0); posteriorly located, considerably away from the anterior extremity of the ilium (1).
399. Ilium, postacetabular process, ventral extension: absent (0) or present, posterior end extends to the level of the ischial peduncle (1).
400. Pubis, ischial peduncle distinct, inset from the proximal end, groove present between the shaft and the peduncle (0) or short, flush with the lateral surface of the pubic shaft (1).
401. Pubis, shaft close to the proximal end, anteroposterior width less than 1.5 times of the mediolateral width (0) or more than 2 times of the mediolateral width (1).
402. Pubis, shaft close to the distal end, anteroposterior width thin anteroposteriorly, with a ridged lateral margin (0) relative thick, without a ridge (1).
403. Pubis, symphysis length more than 80 % (0) or more than 50 % (1) or less than 40 % (2) total pubic length.
404. Pubis, pubic cup absent, posterolateral margin rounded of the pubic shaft (0) or present, posterolateral margin sharply ridged (1).
405. Ischium, shaft minimum anteroposterior width less than (0) or more than 20 % (1) of the ischial length.
406. Femur, thickness sub-equal to (0) or more robust than (1) tibiotarsus.
407. Femur, proximal end, posterior sulcus present (0) or absent (1).
408. Femur, distal end, longitudinal ridge extending proximally from the medial condyle on the posterior margin: present, forming a prominent posterior intercondylar groove (0) or absent, without a distinct groove (1).
409. Femur, distal end, medial condyle transverse width: sub-equal or greater (0); significantly less (1) than the lateral condyle transverse width.
410. Tibiotarsus, lateral cnemial crest, size: prominent (0) or small (1).
411. Tibiotarsus, lateral cnemial crest, orientation: mainly anteriorly directed (0) or mainly laterally directed (1).

412. Tibiotarsus, fibular crest, contact with fibular condyle absent (0), or present, fibular crest continuous with the condyle (1).
413. Quadratojugal, jugal process, shorter or subequal to other quadratojugal processes (0), or much longer than the other processes (1) (Xu 2002).
414. Quadratojugal, ascending process, flattened and well developed (0), or rod-like and very short (1) (Xu 2002).
415. Quadratojugal, posteroventral process, anteroposterior length more than half the length of the jugal process (0), or one third the length of the jugal process (Xu 2002).
416. Antorbital fossa, shape: anteroposterior diameter greater (0) or less (1) than dorsoventral diameter.
417. Antorbital fenestra, size relative to external naris: larger (0) or smaller (1).
418. Jugal, postorbital process, location: considerably anterior to the posterior end of the jugal (0) or nearly at the posterior end so that the quadratojugal process is minimal (1).
419. External mandibular fenestra, size: small (0) or large (1).
420. Dentary, dorsal margin: straight or concave (0) or convex (1) in lateral view.
421. Furcula, cross-section of lateral end: elliptical (0) or L-shaped (1).
422. Ilium, preacetabular process: deep (0) or shallow (1).
423. Dentary, ventral margin: straight or convex (0) or concave (1).
424. Lacrimal, posterodorsal process, orientation: subvertical (0) or posteriorly inclined (1).
425. Anterior caudal vertebrae, transverse processes, distal tapering: absent (0) or present (1).
426. Lacrimal, anterior process, extending anteriorly to interfenestral bar: absent (0) or present (1).
427. External naris, main axis: subhorizontal (0) or subvertical (1) (Osmólska et al. 2004).
428. Parietal nuchal transverse crest delimiting skull roof: present (0) or absent (1) (Osmólska et al. 2004).
429. Procumbent anterior dentary and premaxillary teeth: absent (0) or present (1) New character.

Scoring data
Allosaurus_fragilis
?11000?00000000000100010001000000111011002201?2000000000000
01000000000000001000000000010101001000000001001000000000001?01
000000001????70000000000100010100010100000010000010100000000000
0000011000000010000000000000000000010000000000000000000000000
00000000000000001000000000000000000000000000000000100000000000000
0000000000000000000000000000000000000000000000000000000000[12]00000
1000000000000000000000000000000000000000000000000?00-00[12]?000
Sinraptor
?11000??00?00000001000100010?000000101100020102?0000000000
0010000?00000000?10?00000000101010010000000100100000000001?

0??????????1?00?0?0??00??????????00??10000?0100?00100000000100
0000001100000010000000000000?100?000?00?000000??000000000000
00000000000000000010000000100??00?0??????0?00??00?0????????00??02000
000000010?0000000??0?0000000?00??000??001?0000000000000000000000000
00??????????????????000000000000000000000000?00-00000
Dilong_paradoxus
??100?????000??????00010??10100000?0?0101?0??2?000??1??0?0
??????000?000??0???00000001010?1???00?0101????0????????1????
0?0?0?1?0??????020?10??????0?00000000?1??????0?101?0?1??1?0??
?0?????0??????????100?0??00??0?0?????0?0????0???0?000000000131
0000000000?0???110?0001100?0001011?1??00?10000100010000000001?00010
10?1011000000000010???0?01?00??0?0?0000100?0???000000000000??000000
00??????????????????0?00000?00000000000000000000?00-?0000
Eotyrannus_lengi
?????????????????????0?1?????????????0?0?????????????????
??????00??0?0?????????0000?10??01???00??1?1?????????????????
???????????0?0?0?000??????000?????????????????????????
?????????0???00??????????0?0?????00?????00??00??00?00??
11??0?0?????01?111?00?0?????00????11?0?0??1?0?0??0?0?00???????????
?????????????1?????1??000?0?00?0??00???????????0??000?00?00????
????????????????????000???????0??0-?00?0
Tyrannosaurus_rex
?10000?00?00000002100001101010000000001012010211000010010001
01?0000000000000100100000000101011010000000100101000000101?00
000000??1????0000000010100??0?0100010010001000001011011010000
000001000010001000020000000100000000000001000100000000001
0110000000000010110000000000010100000?00111000100?0?00?0????01001
001010110?0000000???0001000000000000100001000000001000002000001000
00000000000000000000???0-0000?000000?000000000000-002?000
Gorgosaurus_libratus
?10000?00?000?0???10000110111000000011012012 01?21100?0100100
01????000000000001001000000010101101000000??00101000000101?000
00000001????00000000101000010100010010001000001011001010000
0000010000100010000200000001000000000000?01000100000000001
01100000000000101100000000001110000000?01100101000?0?000000??010
0100101011000000000???00010001000000001000010000000010000000000100
000000000000000000000???0-0000?000000?00000000000?00-00000
Tanycolagreus_topwilsoni
??00??????????????????0?0?????0????00000??????????0?00?????
??????????????????00000?????1??0??????????0100000??????0??0?
??????????000000000100??0?01?00010?????????????????1??????100000
0001000000001000000?00000??0?????1??000????00????????0?0???
00??0?0?????????1?????110??00100111100001110001000100010000001?????01
0010110000000000011??1?0?1??00000?00??0?000????0?????00000??00000000
000000000000000000???????0?0?0000000??1???????? ?-000?0

Coelurus_fragilis
??2??000?
??????????????0????1????0000100?010000????????????002????????????0?0??0?00100??
01?0???????????????????1?????1100000000100?0000?1000?0?????00?0????0?0?
?0???000????????????????????????00??????10110??00???111100??1?1????000
1000???????????010?101?0??????????11??1?0?1???00?00??????00????000?000
00000?000000??00000000000000???0?000001000000000??????????1??????
Ornitholestes_hermanni
?0110???0?0?00?1???0?010?011100001?000000?00?100000001?01
011?????00000000000?000000?0101000???00011?1101100000??100??0
010??1?????????????????0?000??????00010001??010?0010?000101?10
000001??0???????00000????00000?0010?00?0?00?000100001000100
10?01??1000?100?000?00001110???????11010?0???????0?????000100??01001
0?0?101?0???????0???000000100000??0??1101??00000000000000000000000000
000000?????00000????00?0???1????0???000000000?00-00000
Compsognathus_longipes
?0110????????????????????1???1??0?????00000???0?01??????00
???????00000010??0???0?0001010101000?010?10?0???0?0?????
01??0?2000001??????0?000?00?10??0???0???0??????????00100?1
00?01?0???????00?0???000000000000011100?00????0?0????0???00
00000011010?0100100001?00?0?0?0?0001100?0001010010????????????????0?0?
10??0100101??1011000000000??0?00?00010?000000??1110?0000??00000000000
001?0000000?00000????00???????0?000????00?????00000000?00-?00?0
Huaxiagnathus_orientalis
?01?0???????????????0001???1??0?0?????00??0?0?0????????????????????00?0010
??????0?000101010?00?0???1???????0??????1?????2?000010???000000000?10??
1010000000020?0??1000100?000001?0?0?0??0?0?0??0?01000000000?011??000????
00?0??0??0??0000000011010?0100100001?0???1?0?00011?0?0000010010000?00000
100010?0000000?0100101??1011000000?0001000?0001000000000?10000?0????
00000000000?0000000?00000?????????0000000????00????????00000?00??0?0
Sinosauropteryx_prima
?01?0???????????????0001???1??0?0?????0000??0???00??010??????
?????000?01?0?????0?000101010?00??01??100???0?0?0?????1??1120
0000100????00000?00?10??1010001000020??????00100?100?01?0?0
0?00?000?0?01000000000?011??0000?000000???0???0000000011010
00100100001?0???0?0?0001101?0001000000000?00000100010000001000001001
010?1?110000000000?10000??0100000000011110000?0???000000000000?00000
00?000000?0?00??00000?00???1?00???00????00???00000?00-?00?0
Deinocheirus_mirificus
???
??1?00000100
??0?10000???0???????
0?0????0???0?0??10?0??110000010
0000000000001?????????????????????1100?????????00???11??1?0??????????
?????????00???000??1?00000000000000?1???????????????????????????????

Harpymimus_okladnikovi

?011???????????????21000???????000?????????0000??00???????????????????
?????020000?0000????1?122????0?0???00?110001?00001?10000?000
00?1??????????0?0?0?0001001001002001 0??00011???0??????????00?
????????????00000010??0000?10110???1?00???000?1?2011001101?
00100?00101?00?0???0000110??00?0??10?000010000010000000000010020????0
????01?0???00000010000?00?1?0000?0000??110?00?0?0000???00000??000000
??001000?????????-?00???????0000?0?????00100?01-?00?0

Pelecanimimus_polyodon

?01???1???1????1?2?21000??10010?????02000?00???????1????00
0??????000?0000?00??0?00021??0001???10011?00????0???????????
?????????0100??02????2?10??00110200??????????????????????????
?????????????????????????001010????1?1????0???1?10?100110020
0100100101?0??00?010001????????0?10?00?11000111 00001000012002??????
??????????????1100??0?0???????011111?0??0?0000??0?0???00000???????0?00?
10??????????0-0?0?????????????????00000??0-?00?0

Shenzhousaurus_orientalis

???????????????????210001?10??0??1???000??00?000?00??????
??????0?00000000??0??0?1?122??1????????????0????0???0???0000
0???0????????????????????????????0200100 00??110?00100?000000
100?00001?0????????????????0??1??1?0011?????0??01020?100?00?100?00?00
????0??????0??1100??????????0?0??0?0?10?0010?0010?0200?0?0?0???????????
???100?0??0?0?00????0???11?0??0?0??0000000000?0??????0??????????????-
?0000?0?100??????????00?00?00-00??0

Archaeornithomimus_asiaticus

??
???????????????????????????????????0001 1000100000 1?100000000
?0??????????010201?20100??0?10100???000?1100011001?00?0010
0000001?00010001000001?3?0000??01??10??1?1????000???????????????
??????????????????1110??00??0?10?0??110?0001000?1??001200??000
10100?11?0?????????0000?000100?00???11??1?001????010?00000000-
0000000?000000000????0???000000?000000?000????????0??0????

Garudimimus_brevipes

?0110????01101????02100110100 1000??0020000000000010?1??0?
0001??1??0000000000020001?1?3??????0110 0???000??0?001??0??
???????????????????????????????????10000001100??????????00?
000000010000?0001000001??0000?01??1110??11?0010???0??1100
11012001001001 01 00000?????1111 0??????????????????????????????
??????0????01011011 0010???00??????000?1000??000???1????0000-
00000000000000??????000???????00????????0000?10?000?00000000000??0-
?0000

Anserimimus_planinychus

??
??0??????????????????0????
??????1???1?2????0001101 00??

00002?3?00????11??10??1?1???
?????????1???0??0?1?0001110001000002200l???????????1?0??????????10
0??????2???00???11??1?0???????1?????00000??000000????1????????????-
??0???????????????000???????????000

Ornithomimus_edmontonicus

?01110?110?101?101021001101?01002110000000000000000100000
00??????000000000010??001?1?3??????001?0011000010000010100
000000?001001?????0112010201000000?101001000000l100011001
0000001000000011000l000l00000203?00001111111101?11100?00
0??0??11011110l200100l00101?0?0??????1?1100?0011011010?01?
0001110001100002200200001010?111?0???10000110000?0002000
00?0011?11100??0000l0000000000?0000000000010000000???00-
0020000?100000?00000000101?00-00000

Struthiomimus_altus

?01110?110??0??101021001101?01002110000000000000001?00000
01?01??000000000l020001?1?3??????001?10l100001000001010000
000001001?????01l2010201000001101001000000l10001100l0000000
00000001l000l000l00000203?00001111111101?111001000??0??110011
01200100l0010100?10?????111l0000010011010001100001l000010000
1100200001010?11110???l0000l10000?0002000 00?0011?1110010?000-
100??00000?0000000000l00000?????00-0020000?100000000000000101?00-
00000

Gallimimus_bullatus

?01110?110110l0l0l021001101?0100211000000000000000l000000
01?0101000000001020001?1?3??????0011001100001000001?10000000000010
0??????0112010201000?01101001000000l10001100l000000l000000011000100
0100000203?0000011111101?1110010000?0??1100110l2001001001010000???
??111100?00?00110l000l100000l0000l00001100200001010l111?0???1000011
00000000200000?0011?11100l0000010000000000?0000000?00010000?00??00-
0020000?100000000000000100?00-00000

Falcarius_utahensis

?01??01000??01?0121?????0???1???????????00??????0?1?00000??????0?
?000??????????001l0000????10?01101?101?000?110011102?011?????????00?1
00011011?0000000100?11001010220001?10101000000010001001010000000?0
0000??0???10?00?0?0111111111?????????????????????00??????0001?11??0?????
11?01?01l1000l000l0000000100021?101????0?0?0?00?000010??0000000000
00?0??0?0?0??????00?00000000??010000??0000000000000000000100??????0000
000???????0000?0????

Beipiaosaurus_inexpectus

???2??1?
????????????0100001????????1????????????????????1?????????0???0???????
?00?0000?100???????????????????????00?0????10??0?000????0??????????
0??????111???1????0?????????????????00?????????????0???0????0100?110001
0?010?0?100?0?1?????11?0????0?1?????0?00???????1???0?0?0??0?0??????00
?00000000???????000???????????????0000??0?????00000?????0101l001??????

Alxasaurus_elesitaiensis
???220?100
??????????0100001??????????0?01?00000?1?001??02?0?1?????????????000?0?
10000000010021?00??100210?0?1?2?????0?????00???????000002?0000??0???1?
?00?0???11??1111????????????????????00???0?011111??????0?100101?011??00
?1??????11001??1?001???00???????00000??0?1??000?010001?0????000????????
0?00000000?0010000??00000000?????00?00000?????0000?000???????1??1??????

Nothronychus_mckinleyi
?????010?????????20??????????????????????????????????00100??????????
????????????001000?????00001101??100001??????1???1??????????0?0????0
1100??????????????????0021?0?110?????????0???010??????0????????????0?
?????0?????0?1111?1??????????????????????????0???101?0???0?0?00???????
?????????????000????001?0??????????????1????11???1????00??????0???????
?0??????????0??01000000000000000000??????????????????????????????????0?
???

Erliansaurus_bellamanus
??
????????????????????????????1???0??????????0?????????????????00?10?
??0?00010????1???1?????????????????0??0??000?000?????????????????0??????
????????11???0?10010?001110000
10100001100001?????01??????????????0100????????1?00??00??0?1?????0??
??0???????????????00000000000000??0000?????0000000?????????????????

Nanshiungosaurus_brevispinus
??
?????????????????????????????0001?00?0??0?????????????????????????????
?????????1?0??0???20020?0?0102??
?????????????2????????????????????????????1?010????????????????????????
??????????10011????????????????????0???0??1???1????????????0???000?
??????????????????????????????0000000000?????????????????0???????

Neimongosaurus_yangi
??21????
0???????????0?000?1?1??10011?0101?0?0?1??2?01000210?1??????000100000
0???????????1??21100??2?????????????00?00000????0??000000?20100?????
0?1???0?0????11?121????0?????????????????0?????01111100 00???010?1?????
?????????????????????????010000?00??0000???????0??0011 101 00????0?01?
???000?000?????10010000000000000000000?????00??????0000000???????1?0
1?0????

Segnosaurus_galbiensis
??2??10
?0000?000???0100001??????????????????????0?????????????????0?10?1?00??
??????0???100211001?200210?0?1020?0???????00??0?00??0000?21?0??00????
1??0?0????111012111???????????????????00?????01???????021?0?0??0??????
?????????????????1000100??0?010011???01??????001??11?01?0???????01???
??0?0000???????010000???000000????0?????0000?100000?00???????11?01??
????

Erlikosaurus_andrewsi
?0110??02?1?1?0???0100111100??10001?100000100011000?0??00
100??11111220010000002000 1?00100001??????????????????????????
???????????????????????0??????????????????????????????????????
??????????????????0?00?021????00??01??00?0?000111?1?11?0001
001?00100100001000000000 1??????????????0??????????????????
???????????????0?0?001100001??????????0??111110 0???1?????010-
0???0??00000??????????????????????????????????????00001011??1-
?0000

Therizinosaurus_cheloniformis
??
??001000?0?00?
?001000?0??0???????
0?0????11???????????????????????????????????????02??0?00??0?011??0001
01?0???1?10????????????????????????????????????11??0???0?10???????????
???????000000000000000000000000??????????????????????????????????

Patagonykus_puertai
??
?????????????????????????????111????????1??2?????????????????0020??3021
0?110??110????????1?????0?????10?00?11?1?00???10011?0??0?????01??0???
1???0??????010???????????????????????????????????1?1??????????00?0?????
??0?????00?1??????????11??1????0???????????????????1???????1????????0
0?????????????000?000000000????????0??110?2??0?0?10??????????????0????

Mononykus_olecranus
??????00???????112??????????????????????????????????????100?
???????????????????????2??0???????01?1?111010011????1??2????
??1???1000?0002000302 10?1122?1100??1?????1??000???????3??0?0
110100111011211000020?00001??00???1??0?0?????01??000?????0?
???????????????????0?111??00?00000000?010001000????00?1?
?????????111??1??00000000???0??00??001100?1???1?00?????010-
?000000000001000000000000000000???000?0??0?2?0?00?10001000-
00?1??0?0?0

Shuvuuia_deserti
?0110100000000?112010000?00??00?111??00001000010101001001
0???1?11000100000021000?021??000??011111101110 1?0??1?201??2
012000100?1000?00020?03021??1122011000?1?1011?2000?00021032
?0??1101001110112110000201000?11000001100000010001?1??0?100
110??0010010010110?00??0?0???????001?00000??0???001?000??0?00
0110??0?????1??111?000?0010011000?00002?001100012111101?0?0?-
01000000000000010000?0000????0000??00??0100?2?0??????01000-00?10-
000?0

Elmisaurus_rarus
??
??
?0??01?1???????????????????????????????????100010??01????0???????

??0??????0000
10000?0??00??????????01?0?0000010??11??????1??0??00??0??1???1?????1????
??

Caenagnathus_collinsi
??20110?
00100?011????3???
???00??0??????
0????0??????????????????????????1??10???????????????????????????????????????
??0???????????0???0??00
00??11?01?????

Rinchenia_mongoliensis
?01?0????0??????????0111?1????1011?10013??00???000?0??????00??1?2211
20?01100??111?1?3??????1??????????1???1???1??0??????1?????????111?1??000
00???1?0??00?10??1??0????????????????????????0????0????0??????0??10??0
01????00???0?????????10101???01?01110?010?10???????????????????1?????????
?????????????????0??????1????????????????????????0?????1???1?????11100
?11???0000??????????????????????????0??????????????0001 0????10?0110

Citipati_Osmólskae
?011001001001??22100011111 10001011?100132100011000001 10011000110
221120?01100101 11?1?3??????1011001 10010111 1001??20???0021001 11?01111
1?010000100?11000001 11002110 0?1 10220?0??1010010?1101000000100?00000
1000000010010011 0?000000000??0???01010110101001110101001?????? 10????
010?00101101001010000000000000000100??????1??101?000000010110010000 0
1??00000100?10?01?1110?0?10000000???0100100000001 ??????? 0?0?000?????
??0??????10011011?010?0110

Ingenia_yanshini
?01?0????????????????1111 1100010?1?1??132??0?1???0?0??0011?????1221
120?01100?0111?1?3?????????0??1??1???1???2???01????2100??? 0011 111111?
000000??10000010100211??11?022000011010?11011010000010010000000000
0?1100?0?11??000?0?0000?0??101010110?01?01110?010?10???????? ??0010?00
1011010010100000010000000110000001011?101?00????0010001 00000000000
00101?10101101 10001 10000000??0010010000000 1000??????0?00000??10000
00000???11011??10?0110

Khaan_mckennai
?01 10?????????????? 00111?110?01011?100132100011 0?000??001 10??????2
1120?01?00?0111?????????? 10110??1??1011110??????1??00210?111001??1111
10000?00?11000001 11002110?1 110220001?101??1??110100?0??100100000000
00000100100110?000???000??0??10101011 0101 0011101010?10?????001 11001
0?00101 10100101000000000000001 10000001011?101100000000001 0001000000
0000000100?10101?0?10101 0000000?210100100000001000000000000000000000
0100000000010011 0?10010?0110

Heyuannia_huangi
??11?????????????????????????????0???????????????0?????????????????1120
?011?0??1?????3??????1????????????????0?2??????????0?111?????11??10?00?1
0??1110?01010021???1??0220??011?1??1???1?010000?0?0????0000000?001??01

??1????00???0????0??1????????????????????0?1?????00?110010?00?01?01000
010000001?0?00?01???00000011?1?1?0?0001000???010?0?00?0000001?1??0?0
1?0????01?000??????01001000??0?1000000000000??000001000000000??????
???1??????

Sinovenator_changii
?0???1002?000011120000101?11100???1?????000011???011?10100?????0
00?001??????0000021??01??????1?1?11100011010000100011 0????1?????????010
111?0??????????000???21??011100200011101100?11110000???00100010 01??
000?????0?1??000011?0??0??001100?10?000??010?1?1100??0?00001111??0010
0?????????????????0?????000000???2011011?101?1???0011??0??1000102?00??00
0??1?0??110???00??000000002101001000010100100002????00011001000000
001???001000001110?0

Mei_long
?0?????????????????11010????00?01?000022100001?10?010??100?????00
010010??1?0000?021??01?00110111100 10?1010?01???011012?111011????101?
11?10?0?1?10000000???2111?1???020??111101110?0111100?0101101?0001?110
11?0?0000???00000????0????0?110011???001?0100101?????????1111??00010
011110?0??0?????0??????00000????????1??1?11100?1?1??000?11??1?21?00000
0?01?010110??0?000000000002101001000010100100002 00?00001100??0?0000
001?????10000011?000

Byronosaurus_jaffei
?????101???101?110011010101 1?00??????20220???????????1?100??????000
0001??11????00021??01?0?0??????010110??????0?0??02????????????????????
??1????0?0??0????????1??????????0??
??0????100???0?00?100?101100??0100??10?00110?????????????????????????
????????????????????????????????11?????????????????0??11?????0????????
??00000??????????????????????????????????00???00?00??01?1000

Sinornithoides_youngi
?0??0???????????????1?000??1???0????0020?2???????1?0??????????????00??
0010???????000110101??????11?1001??????????0011?12102101?????01?11??
0?00??10000000??????????0?200?????1?1???11110?????1????0001?11001?0??
010???0000??00????30110011011??1?0100??1?000?1?0?011??0?00??0011 10
1001010001000000000000??0???11?111110001111000010????21000000000
1?010110??0?0?00000000?1010010?000101 00100?????0?0????????0???00???
?0?0?0?00?1?000

IGM100/44:unnamedtroodontid
??????012??1??0????????????????
1??11???????1?0??????0??
?10000000???01?1?001???00??????
0?0??1???????3?????????????????????1???0????????????????????1?01010 00?
000????00000?????????????1???0?1111????0????????0???0?????0?0??????????
000?????21??????????0???????????00???????0??????????????????????????

Troodon_formosus
???1?1112?1101000001???0?010100??????20220000210?00?1?01100???0?1
0??001?????????011010100???1111100101101111?1000?1020??11??????????

????010????0??????????1??????????????????1??0?011?100?????0100001210?01?
???0?0????0??01000???0?30?????0?1????????10?000???1?10??111?????????1?
01?0????????????????????????????????11?1010111110?????0?101??0??000??0??
???10???01??00000000?1??
?0111???
Saurornithoides_mongoliensis
?01??1?1??1101???0?110001?10000????????2????????????????????1?010?100?
0010??1????0001101010????????1???0??0?????100??1????0????????????????????
??????????????????????002000?0101?10??11110?????????00??211?01?????0??1
?0?????00??00?30?0001101100100000??1?0???1?010??1??0????????????????????
?????????????????0011??1???1??01011110????100?0??0???0????0??01?????0000??10
????00000?????????????????????????????????????0?????00000000??0111000
Zanabazar_junior
?01101?12?110100?001?000??100?000????2022000?21??0????11100??????100
?001???1????00011010100?????????????????1?1000?1020????????????????????
???01???????????????00??0
0???1100??0?30?1001101?001000001010000?110?0????0????????????????????
?????????????????????????1???????????????1??????????????0??01?1??0?0??1????
?00000???????????????????????????????????????00000000??0111000
Unenlagia
???
???????????????????????????????1111?01????0???????1??????????0?0??1?01??
??????????0121101001110201?12??1110011?110?0????010000211?0000??????
?11????0?????0?0?????????????????????????????101??03?0??1??????????
??????????1?0????0211?011???1?0?????211????1?22110210?0001????????11????
????000???????010?10??000???????????????1110?101000???????????0?????
??
Buitreraptor_gonzalezorum
?0110????????????????0?????11?000?????????00000???0011?0????????00??
011???0?????021??10??101101110011??010??01000110 12?121???????0010111
1010?1?100?0????1?110100111 02??112111110??11110?00??10010000211?000?
????0??1?00?0??000000000??1??0?0001001?0?01?0?????0?00111?00310111
120????????????????????????0211011?1?1?0???2?21?????12011021??00010??
1??1?110??0?0000?011111?101011??00101001????1???????1?????1000?000??
?0?0?0??0?0?000
Austroraptor
??11?????????????????0???1100?0?????012?0000?????????????????????0000
011?0?0?????021??101????1011100111 1011?????????????????????????????0
1????????????????????????????????????11?1??0???1?01000???1???10????0
????0?????0????1?200??1??010???????01???0?0???00000111??????????????????
?????????????????????01???????????21????????????????????????????????????
?????11111??????????00??????????????????????1?????????0??00?0???000
Rahonavis_ostromi
???
?????????????????????????????0111111?0?1?11?011012?12?????????0?0??11??0

01????????01111?000111102?001?101??00111?10000??1??100002111000?????0
??1?????0??????0?0????????????????????????????????????1111??3?0???130??????
??????????????????0201?011?101?01112021?????11?1101100??01??????1?11??
???00?000?????01121011??00?11111111110?????11111211011111??????????0??
?0????

Archaeopteryx_lithographica
101?0010??000??112011010??11100?011?00022100010?1000111?100111?10
000001100002?000[01]020??[01]0100?1?0??100?1??1?0?0???00210121121?0??
???001011110000?110000001011[12]1?0001?102000121021101?111100000?10
01000021110?0000001001100000?0?00001020010111010001001001 01?0??00?0
001?11110?3?0121130110?01[01]001000100?010010202010011?101101112011
00000100011210000000000110101?0000000000000000?10211111011001??011??
?00101011?002110??????11100100100100000

Wellnhoferia_grandis
?0?????????????????110?0???????????????????????????????????????00??0
11???????0[01]020??0010???????????????????0????02?00231???0?????010111
10?0????0?0000001??1????1??0??0?1?1?21?01?1?1100?0????01000???110?000
??010????00?0???1000?0200101?1??????0?00???0???0?0?0?0??11?1?0310?2113
0?10?011001000100?010010202???011?1?110111201?000001??0112100000000
01?0101??????0?00000000?1021111101 1001??011???00101011?002110??????
???00100?00??????

Jeholornis_prima
10???????????????????1?????0????????0??2???0??????????????????????120?0
00??00200?1?1?2??0???0?????????????1???1????02???211?11???1?1?0103?110
?00??11000000012?1??0???10?0???????21?0??????1?000??1?01010?21110?000?0
010????1000??00?0102?0?0?1????0?10?1???01?01?0???01?1111?0310121131
11??011000?11001100000000 2???01??1?1?011120?0100101??01111000000000
?10101?0??1??00000000000111 21?11??110111??111???1?11111?1020?0???1?01
111?011011???000

Sapeornis_chaoyangensis
?011????????????????11011?????00001?000002?00011???00?????????????000
?010??00?00000??3?????0????????0??0?0010?2???0??????21?1111????1000111
10?00??1120000?011?1??01??100?0?1?1021101???010000??0??1110010??00
0?0010?1???000???1010?0??0?0?1101?00100?0??01?0??0?????1?111?00310121
13111?1011100?1110011100?1?02???011?1?110111????1??101?00?01?0000000
00110?0110??1?00?0000000020121111?0110111??111?1?1?111111102?01?????
?111110?10110?0100

Sinornithosaurus_millenii
?011?????0?????????0????11100????10002?1100?1?1?0??1???????????00?
00110??12???0?01011[01]01????????1??1??0???0???00?1??1??????0101?0103
1110?00???0?0000010?211?0??10?2?0?1[12]1022?0????1???????1?0100002111
0?0?000?1001??0000???0??00?0001001101300?0?100101??[01]00110?0?????0
0010111121?00101100001000111100001?211111??1?100001201?0000012??112
?000001000110101?0????0??0000000101010110111010100100002 00?00001000
1000??????11100100000100000

Microraptor_zhaoianus
1?1???????????????????00??0??01
1?0?1????0?0100000?????001?1??01?1010?01???0110?11?2111?01?100?1?1?10
100?110100001101211 0?011002?0?1[12]1022100?11111??00?1001?10121110??
00??010?11?000????0??00?00?101?1???????????00??1???0??0??0?111100010011
12110??0110000100011110010002111110?1?100001201110000020?112100000
10001?0101?0??000?0000000000010111111 0101?010000200?00001000100000
?00011100100000?0?0?0

NGMC91:unnameddromaeosaurid
?011???????????????100????11??00?????001???0????????????????????000
0110??1??0?0?010??10??0?????????????????????01???1????1?????10??031?10
?00??1?100001????????????????????????????1???0?????????????11????0???10
????000????10????00010111013?01?0100??1???0?1?0?00?????0?0?0011121110?
?011000?10001101001 00?????1??1?1??001201110000??????21?000010001101
0??0??????0000?????010101111110101001 0000200?00001000100 0??????11100
100000?0?0?0

Confuciusornis_sanctus
10110???????????????11000?0???00001???0?12?00?01??0?0??0???01??
?????000010010?10?0001?1?3?????0???1??1???1011?0?2????0???????2??
111?110100013111400011111000010111101????100?0012103210111??1
0??11?112111101101 0?000000100110?000???001??0??0011????3001000
00?01?0?10?????11111?000101211211101011000010101011001 0002???0
11?1?11011120001011 01??0011?000000000?10?01?0??100?20000000?0-
111010?0110111?01111?01111211?1?2101???11?111110110?0???010

Protopteryx_fengningensis
10?????????????????110?????????????0???????????01???????????000??
1???????0?0??2???0??0????????0?????1??2???????????2??111?11?11??031?100
00??1120000001??????????????0?1?1??????????1???1????????1111010?0?0??010
1????00?????0????2?01?111???00?10?00?01?0??????01?????1?3?01211311 11??
011000?110010110001??????11?1?1?0111200?0??10??????1??0000000011 0?0
1?0???0??0000000???121?111011011???11??????11??????2?01???1?0111??0?0
?00??????

Yanornis_martini
?01???????????????111?0??????0??????0??????????0?1?0??????
?????000?011??00????00020??0000????11?????????0?2???????????
??111?1???0?0031?10?00??1120?00???????????????00???121?32?01??
?0100?1??11??112?01010?000??010????00?????101?1?200001?1???0
0110?00??1?00?00?0001111??1???012113111??0110?0011?0?11111
01??????01??1?1101??20000??0??????1??00000000110?0??0??10??-
?0000000???1?????1011011?1?111??0??11??????2??1?????????11-00?-0??????

Bambiraptor_feinbergi
?0110?????0?0?1??1??00?0?01110000??10?01??10?1????0?01?0?????????00
??010??1??0000?01011100?????0?120011??10?0???0011??110????001100101
1110?001?1010000010?21?001110020001?1021100?1?110000??10010001101?0
?000?00?0?1??0000????000?0000000?1013001??100101?10?011100011011000?

01111201011010001000001011001010201101101?100001001100?00120010110
00001000010111110???000000000000??010110111010100100002 10?00?0100010
00000000000001?0000100000

Tsaagan

?011001001000012011200001010100001?1000121101 10001000??00001?0??
?00000100?11110001010101000????0??2??????????????????????????????????
???0
??00??00?0?0000???0?0??000100 13?0001000001?1???0?0?0?????????????????
??0???1?????0
??????????00000???00000000??0??
?000

Adasaurus_mongoliensis

??1100?????????????????????????0?????001???????????0????0?????????????
??0?????0?0?0????????01???0?2?01?1?????011000????1?????????????0??1?1????
???????????002211001110???0????021?0??1???00????????001100?0010????0
?1??000???????0?000??????????0???????1????0?????01111??00?0???????????
??????????????????02???011?10110001011 10???12???01?00??01??????1???0??
??0?0000?????????????????????????????????01000100000?????????????0?????
??

Velociraptor_mongoliensis

?011001001000012011200001011100001110001211 1?100010001000011 11010
10000010011110000101011100001100012001111101111100011011 11021110011
10011111101001110000000100221100111002000111021100111 10000000100100
01101001000000?0011000000000000000000001001300001000001010010 10000
110100001001111010010100010000010110010102101011?001100000111 00000
1200101100000100001010110000?000000[12]0000110101100000101001000021
?0?01010001000000000000000?00100[12]0000

Saurornitholestes_langstoni

??????????????????????????01110???????????????111?????????????????11????0???01
0?????00???01011[01]000?11000120011111011011100?1011??11??????????01?1?
??????10??000100221?0?11??????????????????????????????????????0???00?0?
???00?000?0???0000?????013?????0??00??10?1?1000011????00100?????1??????
??????????11001??0???????????????????1??00?1??010??0?????0?0?0?01?0??0?
?????????0???000

Deinonychus_antirrhopus

?0110????1???????11?0000?011100001110001 2??????0?1????0?00?11010?00
0?0100111?1000?010 11000?110001 2001?11011 1???00110110?21?10???1?010
1111010011 10000001 0022110011 100200011 10?010011 1110000000100 10001101
000000?00?001 100000?0?000?0000??00?0013000?000???? 000?1?1 10001001???
???011 1101001010001000000011 00102000101 1000110001011 10000012?0000
1000001000010101 10000000000000000?1010???1000001 001 00002 10000100??0
??0???000000000000?00100000

Achillobator_giganticus

?????????????????????????01110???
??????????000101?????0?01200?111100?????????011??1?????????0101????????

?????00?010220?0111100100?011010?00?11110?0???00?000?101????0??????1
???0????????0?000?????013????????????????10??00000??00?0???1??????????
?0?????01?0????0000100100?1??????11??????1000000?0?????????0??1?????0?
??0??0000011010?1?????????????????0?????0100?000000?000?????????????0??

  Dromaeosaurus_albertensis
?0??001000000000010??0?0?0??0??0??01110????1011????10001?1001100???00
0001001111000?000101000???
???1??????00??????
00?0?0000???0?00?000?00????0??00?0010010100000????????????????????????
????????????????????????????0111?????????????????0??01?????000???0??
?00000??00000??0???000

  Utahraptor_ostrommaysi
??????????????????0?0??????????00????????????????????????????????
????????01???101??0????00??1011111?0??????0?1011?????????????01??????
??????????????2?????11001??0???0??????1???000???00?0??????????????
???0??????0???????0?0?????????????????0?????000????????????????????
??????????????001?00??????????????????10???0??????????0??????????0???
??????0???????????????????????????????????000??????????????0??

  Atrociraptor_marshalli
??????????????????????1110????????????????????????????????????0?1
????1????01000111?00???
??0?
?????0??0?00?00?00?00?3??????0?????010?0110????????????????????????
??0
0000???0?0

  Epidendrosaurus_ningchengensis
?0??????????????????????????10000?1?00?????????0????????2??00
000?002000?????????????????00?010?00??01?01011????1??????1011?10
?00??1010000011?1???1??020?0?0?1?02?0??1?0100?0???0????000010?000?0
110?1??100???00?0?0?0??0??????????1??????10???10???00?1111????00210201
0??01000100000000000000????011?1?1?0110200001100??00011000000010?
11101?0??1?000111???????0?0000001010????????0???10???????0??????????
?11???0?0?1?

  Epidexipteryx_hui
0???????????????0??????????00??0000??00?0011????000???????????11
0?0?0????00001020--10000????????0???0?0?1???102??123?-
100-1????0101?01000????????00??0011?10?????020-
0?011?02?0????????0????01111??0?10??00???10??1?1-
?0?1??000???110??01?????????????10?1???0-
1?00?111???0?0021120??????1?0???????000000??00---
01??10110????????????10?000110?000?0??1?1??11???100?0111000002-0?0000
000110???0???????????00????000?????????11011?11???111

  Hagryphus_giganteus
???
???

10000001??0????????
???0?01110000000
10001?00100?????????????????????0?11????????????????0??0?0??????????????
??

Alvarezsaurus_calvoi
???
????????????????????????1???100?????????2?1??2?12????????????0?0?0???????
??????1???00111?0??2?????????????????1?00???0?01??10000000???0?????????1?
??0??????????0????????????????????????????????111110?0?????????????????????
???????0?1????0??????????1?????000?0???????0??1000???0?????11??1????0000?
????????01010000?????????????????0??????????????????????????????0????

Microvenator_celer
??21?10?
0????????????????????011?0?11000111100??0???1002????????????1000??0010
00??????0??1002???0111????0?????10110?0101100?0010010??????0?0??0???0??
?11??0?0????00??0????????????????????????00???????0?111?????0011110??0???
?????????????0?0??0???0??011????????????????????0????00000?1????0????????
0??1?????????????????00001000??????????????0?110?0?0?000?????11?01?0??
??

Avimimus_portentosus
?01?0???10001?00??0??1?1????????1???1?????00?11??0?00100110??????2?1?
??00?0??0110???3???????011011110101?0?00??100????????????????11?00010
0???110?????00?211?01??0?2000??0101000011000?0???10011110020??00?0??
?01?0??00000?00??0?????0???????1?????1011???0?????001?????????111??1?????
?0????????????????00??011?101?0??00000????1000002?0000?0?????????10??0
11?0000??????????????00001?????????????????0?100???0?00?????????1????110

Caudipteryx
00110??????????????0?110?010000011?00013210001?00?00????????????21
010?00?????10031?3?????00????0?100???????0??01????210?111000??001010
001?00??1000000010020??11?02200?111?10?10?11?01??0???0100000200000
0000010?110?000??0?00?0??1000100100110100111?0010????001110?10?00
01110100?01000000010?0001101?00001011?1011000000000??0100000100000
00000110101?0??000010000000??001001??000001??00????000000000?1000??
000???00000?00??0111

Oviraptor_philoceratops
?01?0??????????1??00?????111?0?011???0132??0?1?0???0?100?????1????11
20?01100?01?1?1?3?????0?1???1??????1?0?????????????????1?????11?1?0?0
00001?110?0001???21?00??00100
100?????00?0000??????0?01?1?010??1??1?00?010?10?????0?111?010?00?11?01?
?1010000000000000100?????????????????????1000?????????00??100?10?0
1?0?10???1?000000?0010010??00001000????000000000??10000000000000110
11?010?0110

Conchoraptor_gracilis
?01?0??????????1???00111?1100?1011?1001321000110?000?????1?0????2211
20?010?0?0111?1?3??????????????010??1?012?110??1???0?0111???1?11?110000

10????????011?00210001?102???0???01001??11010?0????0??000000??0?0??00??
0110??0?000000??0???0101011?101001110101????????????????0?????1?1????010?
0???0?00?0000111000??????1???110??????0?00001000000????0???0?10?0??0110
101?0000000???????????????????????????????0????????????????00011011??1000110

Chirostenotes_pergracilis

????????01?01101??0???1?11??0??????????????????????????0
1110???0??????????????????1??????????????1101?11????1?1
2????????????????????101????????????000110021?0001?0220001110?00?0???
??0?????10?100?00100000????0???1??0?0?00???????????????????????????1??
???????1?1???????0??????00?????000010?0?1001000?001?1??101?000000010
001110000?1?000000??0??0????????001???????????????????????????????????
????0?1?0?????????????????[0123]1?[12]???0

Incisivosaurus_gauthieri

?01 1 0 ? ? ? 1 ? ? 0 1 ? ? ? ? ? 1 0 0 0 1 0 1 0 1 0 0 0 1 0 0 1 ? 0
?01 0 ? ? 0 0 0 1 1 0 0 0 ? 0 1 0 1 0 1 1 0 1 0 1 1 1 1 2 0 0 1 0 0 0 0 1
001?1?02022??0002???
???00??00??00?0?
?000???0010?00010?10?01001001011 0010??001???????????????????????????
???0??11?????0110??0???
0000??---11??10?0111

IGM100/42:unnamedoviraptorid

?0110??????????????0011 11 110?01011?1001321 00?11??000?
??01100?111221120?01100??111?1?3??????10??0???1??1?1?1??
?1??01??00?10?1???01?1111110000?0???10000011002110?1110220??1?1010
?1??1?01000???????0000100000001 00?001?0??00?0?0?00?0??1010101101010
01110101??10????001010010?001011?100?010001?00000000001000001011?
10110??00000110010?0001000100 0100?10101111101011????????????????????
???1??????

Anchiornis_huxleyi

001100?????????????11010??111?00????0?2?000??1????1?1??????????-000
??01??1?????00020--01??0?101??-00?0??010?01???011012112?011----
011131110000??1000000010?21?1001102200?1?112112???1?1?000??1?01?
0012211 00010?1000?1??00?0???0????030110?1101000100100?01?0???0-0?
011111101?0001112011010100010101001110010102001011?1011000?20?
1100001000?121000?00000110101?0?0??000000000000002101011110111 01??
011?1??1?110011?12?10???10????00100000110 0?0

Xiaotingia_zhengi

?01???????????????????1?1????1111000???0?1??????????01?1??????----
-?001001??1???00??022?-01??0?-0???1?0????000?01?????1???---10?1----
01?03101?10???10200001011?1110?12012?????103?12??????????10????????1?
010001010000?0??0?10??0???102?110111?1??0110100?01?1100?-0?0?1111?
013?1?1111211001010001??0001011001020201????0??1??
00100011100012?0?1??000?00000??0??1????00?0001100
0002?11?11110010?0???????????????020 1 1?????10???10????
0?10010011?0?0

Appendix 2
Re-Scored Data from Xu et al. (2008); Hu et al. (2009), and Agnolín and Novas (2011) Data Matrixes

Character numbers from the original analysis are given at left, and denoted by the letter "H", whereas scores favoured by our current analysis are denoted by the letter "A" (present paper): **Ch. 20:** *Epidexipteryx*: H = ?, A = 0 **Ch. 53:** *Sinornithosaurus*: H = 0, A = ?. **Ch. 71:** *Wellnhoferia* H = 0, A = 1; *Archaeopteryx* H = 0, A = 1; *Buitreraptor* H = 0, A = 1; *Austroraptor* H = 0, A = 1. **Ch. 82:** *Wellnhoferia* H = 0, A = 0/1; *Archaeopteryx* H = 0, A = 0/1; 90: *Sinornithosaurus* H = 0, A = 1; **107:** *Archaeopteryx* H = 1, A = ?; *Confuciusornis* H = 1, A = ?; **125:** *Wellnhoferia* H = 0, A = ?; *Archaeopteryx* H = 0, A = ?; **130:** *Sinornithosaurus* H = 1, A = 0; **172:** *Sinornithosaurus* H = 1, A = 1/2; *Microraptor* H = 1, A = 1/2; **175:** *Wellnhoferia* H = 1, A = 2; *Archaeopteryx* H = 1, A = 2; **235:** *Microraptor* H = 0, A = 1; NGMC91 H = 0, A = 1; *Sinornithosaurus* H = 0, A = ?; **Ch. 240:** *Confuciusornis* H = ?, A = 3; **Ch. 245:** NGMC91 H = 1, A = 0; *Sinornithosaurus* H = 1, A = ?; **Ch. 250:** *Epidexipteryx* H = 0, A = 1; *Epidendrosaurus* H = 0, A = 1; **Ch. 251:** *Epidexipteryx* H = 1, A = 0; *Epidendrosaurus* H = 1, A = 0; **Ch. 253:** *Microraptor* H = 1, A = ?; NGMC91 H = 1, A = ?; *Sinornithosaurus* H = 1, A = 0/1; **Ch. 322:** *Wellnhoferia* H = ?, A = 1; *Archaeopteryx* H = ?, A = 1; **Ch. 330:** *Unenlagia* H = ?, A = 1; *Buitreraptor* H = ?, A = 1.

F. L. Agnolín and F. E. Novas, *Avian Ancestors,* SpringerBriefs in Earth System Sciences,   93
DOI: 10.1007/978-94-007-5637-3, © The Author(s) 2013

Appendix 3
Character Diagnoses for Selected Clades

- Scansoriopterygidae + Oviraptorosauria: 20.0, 22.1, 33.1, 39.1, **58.1**, **68.1**, 69.1, 74.1, **79.1**, 85.3, 252.1, 255.1, 263.0, 269.1, 288.0, **355.1**, 416.1, 427.1, 428.1.
- Paraves: 75.1, **76.1**, 138.1, 141.1, 144.1, 183.1, **201.1**, 322.1, 323.1, **333.1**.
- Dromaeosauridae + Averaptora: 71.1, 103.1, 175.2, **180.1**, 296.1, 297.1, 300.1, 319.1, **330.2**, 342.1, 374.1, 377.1.
- Dromaeosauridae: 18.1, 53.0, **99.2**, 106.1, 130.1, 198.1, 233.0, 253.1, 346.0, 392.1.
- Averaptora: 90.1, 162.0, 200.1, **290.1**, 295.1, 320.2.
- Microraptoria: 154.1, 176.2, 257.1, 278.1, 288.0, **308.1**, 315.0, 378.1.
- Unenlagiidae + Avialae: 76.0, 159.0, **270.3**, 391.1, 400.1, **405.1**.
- Unenlagiidae: 85.1, 95.1, 175.1, 263.0, 332.1, 364.1, **365.1**, **366.1**, **367.1**, 368.1.
- *Austroraptor + Unenlagia*: 381.0.
- (*Anchiornis + Xiaotingia*) + Avialae: 21.1, 141.0, 262.1, 330.0, 342.0, 376.1, 380.1, 386.0, 403.2, 410.1.
- *Anchiornis + Xiaotingia*: **52.1**, 71.0, **89.1**, 125.0, 133.1, **178.2**, 206.1, 232.1, 269.1, **369.2**.
- Avialae: 75.0, 106.1, 274.2, **277.3**, 317.1, 318.1, **372.2**, **373.1**, 381.0.
- Archaeopterygidae (*Archaeopteryx + Wellnhoferia*): 126.0, 290.0, 324.0.
- Ornithurae: 110.1, 335.1, 371.1, 382.1, **401.1**.
- *Jeholornis* + Avebrevicauda: 196.1, 278.1, 323.0.
- Avebrevicauda: 110.2, **121.2**, 134.0, 148.2, 195.1, 197.1, 201.0, 406.1.
- Pygostylia: 175.3, 176.2, 188.1, 192.1.
- Ornithothoraces: 52.1, 244.1, **268.1**, 324.0.

F. L. Agnolín and F. E. Novas, *Avian Ancestors,* SpringerBriefs in Earth System Sciences,   95
DOI: 10.1007/978-94-007-5637-3, © The Author(s) 2013

References

Agnolín FL, Novas FE (2011) Unenlagiid Theropods: Are They Members of Dromaeosauridae (Theropoda, Maniraptora). An Acad Bras Ciênc 83:117–162

Gianechini FA, Apesteguía S, Makovicky PJ (2009) The unusual dentition of *Buitreraptor gonzalezorum* (Theropoda, Dromaeosauridae), from Patagonia, Argentina: new insights on the unenlagine teeth. Ameghiniana 52:36A

Hu D, Hou L, Zhang L, Xu X (2009) A pre-*Archaeopteryx* Troodontid Theropod from China with long feathers on the metatarsus. Nature 461:640–643

Novas FE, Pol D, Canale JI, Porfiri JD, Calvo JO (2009) A bizarre Cretaceous Theropod Dinosaur from Patagonia and the evolution of Gondwanan *dromaeosaurids*. Proc Roy Soc London B126:1101–1107

Osmólska H, Currie PJ, Barsbold R (2004) Oviraptorosauria. In: Weishampel DB, Dodson P, Osmolska H (eds) The Dinosauria 2nd edn. University of California Press, Berkeley, pp 165–183

Pérez-Moreno BP, Sánz JL, Buscalioni AD, Moratalla JJ, Ortega FJ, and Raskin-Gutman D. (1994) A unique multitoothed ornithomimosaur dinosaur from the Lower Cretaceous of Sapain. Nature 370:363-367.

Xu X (2002) Deinonychosaurian fossils from the Jehol Group of western Liaoning and the Coelurosaurian evolution. Dissertation for the Doctoral Degree, Chinese Academy of Sciences, Beijing

Xu X, Zhao Q, Norell MA, Sullivan C, Hone D, Erickson PG, Wang X, Han F, Guo Y (2008) A new feathered Dinosaur fossil that fills a morphological gap in avian origin. Chin Sci Bull 54:430–435

Zheng X, Xu X, You H, Zhao Q, Dong Z (2009) A short-armed Dromaeosaurid from the Jehol Group of China with implications for early Dromaeosaurid evolution. Proc Roy Soc London B 277:211–217